Polarography and Other Voltammetric Methods

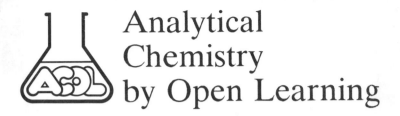

Analytical Chemistry by Open Learning

Project Director
BRIAN R CURRELL
Thames Polytechnic

Project Manager
JOHN W JAMES
Consultant

Project Advisors
ANTHONY D ASHMORE
Royal Society of Chemistry

DAVE W PARK
Consultant

Administrative Editor
NORMA CHADWICK
Thames Polytechnic

Editorial Board
NORMAN B CHAPMAN
Emeritus Professor,
University of Hull

BRIAN R CURRELL
Thames Polytechnic

ARTHUR M JAMES
Emeritus Professor,
University of London

DAVID KEALEY
Kingston Polytechnic

DAVID J MOWTHORPE
Sheffield City Polytechnic

ANTHONY C NORRIS
Portsmouth Polytechnic

F ELIZABETH PRICHARD
Royal Holloway and Bedford
New College

Titles in Series:

Polarography and Other Voltammetric Methods

Analytical Chemistry by Open Learning

Authors:
TOM RILEY   ARTHUR WATSON
Brighton Polytechnic   *Paisley College of Technology*

Editor:
ARTHUR M. JAMES

on behalf of ACOL

Published on behalf of ACOL, Thames Polytechnic, London
by
JOHN WILEY & SONS
Chichester · New York · Brisbane · Toronto · Singapore

Published by permission of the Controller of
Her Majesty's Stationery Office

British Library Cataloguing in Publication Data:

Riley, Tom
    Polarography and other voltammetric methods.
    —(Analytical chemistry).
    1. Voltammetry
    I. Title   II. Watson, Arthur   III. James,
    Arthur M.   IV. ACOL   V. Series
    543'.087   QD115

    ISBN 0 471 91394 4
    ISBN 0 471 91395 2 Pbk

Printed and bound in Great Britain

Analytical Chemistry

This series of texts is a result of an initiative by the Committee of Heads of Polytechnic Chemistry Departments in the United Kingdom. A project team based at Thames Polytechnic using funds available from the Manpower Services Commission 'Open Tech' Project has organised and managed the development of the material suitable for use by 'Distance Learners'. The contents of the various units have been identified, planned and written almost exclusively by groups of polytechnic staff, who are both expert in the subject area and are currently teaching in analytical chemistry.

The texts are for those interested in the basics of analytical chemistry and instrumental techniques who wish to study in a more flexible way than traditional institute attendance or to augment such attendance. A series of these units may be used by those undertaking courses leading to BTEC (levels IV and V), Royal Society of Chemistry (Certificates of Applied Chemistry) or other qualifications. The level is thus that of Senior Technician.

It is emphasised however that whilst the theoretical aspects of analytical chemistry can be studied in this way there is no substitute for the laboratory to learn the associated practical skills. In the U.K. there are nominated Polytechnics, Colleges and other Institutions who offer tutorial and practical support to achieve the practical objectives identified within each text. It is expected that many institutions worldwide will also provide such support.

The project will continue at Thames Polytechnic to support these 'Open Learning Texts', to continually refresh and update the material and to extend its coverage.

Further information about nominated support centres, the material or open learning techniques may be obtained from the project office at Thames Polytechnic, ACOL, Wellington St., Woolwich, London, SE18 6PF.

How to Use an Open Learning Text

Open learning texts are designed as a convenient and flexible way of studying for people who, for a variety of reasons cannot use conventional education courses. You will learn from this text the principles of one subject in Analytical Chemistry, but only by putting this knowledge into practice, under professional supervision, will you gain a full understanding of the analytical techniques described.

To achieve the full benefit from an open learning text you need to plan your place and time of study.

● Find the most suitable place to study where you can work without disturbance.

● If you have a tutor supervising your study discuss with him, or her, the date by which you should have completed this text.

● Some people study perfectly well in irregular bursts, however most students find that setting aside a certain number of hours each day is the most satisfactory method. It is for you to decide which pattern of study suits you best.

● If you decide to study for several hours at once, take short breaks of five or ten minutes every half hour or so. You will find that this method maintains a higher overall level of concentration.

Before you begin a detailed reading of the text, familiarise yourself with the general layout of the material. Have a look at the course contents list at the front of the book and flip through the pages to get a general impression of the way the subject is dealt with. You will find that there is space on the pages to make comments alongside the text as you study—your own notes for highlighting points that you feel are particularly important. Indicate in the margin the points you would like to discuss further with a tutor or fellow student. When you come to revise, these personal study notes will be very useful.

Π   When you find a paragraph in the text marked with a symbol such as is shown here, this is where you get involved. At this point you are directed to do things: draw graphs, answer questions, perform calculations, etc. Do make an attempt at these activities. If necessary cover the succeeding response with a piece of paper until you are ready to read on. This is an opportunity for you to learn by participating in the subject and although the text continues by discussing your response, there is no better way to learn than by working things out for yourself.

We have introduced self assessment questions (SAQ) at appropriate places in the text. These SAQs provide for you a way of finding out if you understand what you have just been studying. There is space on the page for your answer and for any comments you want to add after reading the author's response. You will find the author's response to each SAQ at the end of the text. Compare what you have written with the response provided and read the discussion and advice.

At intervals in the text you will find a Summary and List of Objectives. The Summary will emphasise the important points covered by the material you have just read and the Objectives will give you a checklist of tasks you should then be able to achieve.

You can revise the Unit, perhaps for a formal examination, by re-reading the Summary and the Objectives, and by working through some of the SAQs. This should quickly alert you to areas of the text that need further study.

At the end of the book you will find for reference lists of commonly used scientific symbols and values, units of measurement and also a periodic table.

Contents

Study Guide

In recent years electrochemistry has undergone a revival, particularly in its analytical applications. Gone are the days when the main applications centred around such techniques as the measurement of pH values or the polarographic analysis of mixtures of metal ions. Not only has the accuracy of these methods been drastically improved but exciting advances have resulted in the establishment of new techniques. This has been helped by advances in instrumentation and electronic circuitry. New polarographic techniques, some with detection limits of 10^{-8} to 10^{-9} mol dm^{-3}, have been developed and applied to the determination of oxygen in gas mixtures, biological fluids and water in addition to metal ions. Polarographic methods are now widely used in the analysis of organic compounds including naturally occurring mycotoxins (eg fungal metabolites), compounds used to improve food yields (eg insecticides and fungicides) and environmental pollutants (eg nitrosamines, detergents, azo dyes).

Such non-destructive methods, with rapid response times and low detection levels are now in competition with other methods of analysis, such as spectrophotometry and AAS.

This Unit is designed to introduce you to a wide range of voltammetric techniques. It is intended to impart sufficient knowledge to enable you to understand the basic theory, the practical aspects and the scope of each technique and to enable you to relate the techniques *via* features they have in common such as electrode and solution behaviour and electrical circuitry. You will not become a practicing expert in any one of these methods – quickly, this will require more study and certainly more practical experience. On completion of this Unit you should be able to select a suitable method for a particular application and you will be prepared for further more detailed study of individual techniques. A good basic understanding of how the technique works is necessary in order to be able to use it in an intelligent manner, for example in the choice of operating parameters. In real life applications in the laboratory, you will have to use publications in the literature, and this Unit will introduce you to the terminology that you will encounter.

In someways it is even more important to know when and why a method will not work. You will learn that in general only diffusion controlled polarographic waves are of analytical value. The other types of process are discussed, not only out of any theoretical interest, but so that they can be identified and avoided in practice. The all important analytical application must always be in mind and how it might be affected by the material under study at the moment.

The text contains many examples of the electrochemical behaviour of compounds or classes of compounds. These are included as illustrations to help your understanding of the general principles and to relate those principles to some real compounds. It is not intended that you should try to remember any details of these examples, it is the principle that counts. Working in a real laboratory you will build up your own experience of a range of compounds that would probably be more useful to you in that environment than the examples included here.

To achieve these objectives you are first introduced to some of the basic definitions, conventions, theoretical principles and practical approaches of solution electrochemistry. Only those aspects of the subject relevant to subsequent studies of voltammetric analytical techniques are covered. The remainder of the Unit is then devoted to a discussion of these analytical methods.

It is assumed that you have attained at least level III in Chemistry (BTEC) or GCE 'A'-level with a good understanding of the Physical Chemistry components. In addition, some study of Physics up to 'O'-level or level II (BTEC) would be of benefit in understanding ions, their associated electrical fields, coulombic forces and electrical conduction.

SI units are used extensively, others are used as necessary. Some tables showing the relationship between the various units are provided.

Many texts are quoted as supplementary reading for this Unit. Unfortunately, no one book covers all the aspects dealt with here. As a general introduction the books by Fifield and Kealey, and Crow and Westwood are useful.

You are reminded that other Units in the ACOL scheme are devoted to other electro-analytical techniques. Those available at the moment are: *Principles of Electroanalytical Methods* (some of the material overlaps with this Unit); *Potentiometry and Ion-selective Electrodes*.

It is hoped that as a result of studying this Unit you will be encouraged to study further in the area of electro-analytical chemistry. If you come to realise the potential and the advantages of some of these techniques and become a user then even better.

Supporting Practical Work

1. GENERAL CONSIDERATIONS

Experimental facilities for dc polarography are available in most laboratories. It is strongly recommended that three-electrode circuitry is used for even the simplest experiment. Potentiostat circuits capable of dc mode experiments can be made cheaply and it is counterproductive to use old fashioned two-electrode equipment. The more advanced forms of polarography and stripping voltammetry do require that commercial apparatus be available but one piece of equipment can often be used for all the experiments. The experiments suggested are designed for the most part to occupy a three hour laboratory period.

2. AIMS

There are three principal aims.

(*a*) To provide a good grounding in voltammetric techniques by means of experiments designed to illustrate the principles of dc polarography.

(*b*) To give experience in the assembly of apparatus using three-electrode circuitry and in the choice of electrodes, solvents, supporting electrolytes for a particular analysis.

(*c*) To give experience of realistic analyses using the following techniques:

    (*i*)   dc polarography;

    (*ii*)   normal pulse polarography;

(*iii*) differential pulse polarography;

(*iv*) anodic stripping voltammetry.

3. SUGGESTED EXPERIMENTS

1. Using pure chemicals

(*a*) Use a 3-electrode dc polarograph to produce polarograms for Cd(II) and Zn(II) in aqueous 0.1 mol dm^{-3} KCl. Investigate the effect of:

(*i*) deoxygenation;

(*ii*) maximum suppressor;

(*iii*) height of mercury column;

(*iv*) change of supporting electrolyte, eg $(C_2H_5)_4NBF_4$.

(*b*) Analyse an approximately 10^{-4} mol dm^{-3} aqueous solution of Cd(II) using dc polarography.

(*c*) Compare the dc, normal pulse, differential pulse methods by analysing an approximately 10^{-4} mol dm^{-3} aqueous solution of Cd(II). Use this experiment also to investigate the effect of instrument variables on preformance, eg pulse amplitude, scan rate, drop time.

(*d*) Determine Pb(II) present at the ppb level in an aqueous solution of Pb(NO$_3$)$_2$ using anodic stripping voltammetry at a mercury film or hanging mercury drop electrode.

2. Realistic Samples

These experiments would take more than one 3-hour period. The choice would have to be left to the individual centre but examples are:

(*e*) heavy metals from an effluent sample.

(*f*) drug from a body fluid.

The student would be expected to use an appropriate separation method before selecting the best voltammetric technique and applying it using a suitable calibration method.

Bibliography

1. 'STANDARD' ANALYTICAL CHEMISTRY TEXTBOOKS

Virtually all the books giving a comprehensive treatment of (instrumental) analytical chemistry contain one or more chapters on electro-analytical methods of analysis. An example is:

F W Fifield and D Kealey, *Principles and Practice of Analytical Chemistry*, International Textbook Co Ltd, 2nd Edn, 1983. A survey of analytical chemistry introducing polarography in the context of other analytical methods.

2. ELECTROCHEMISTRY TEXTBOOKS

(*a*) D T Sawyer and J L Roberts, *Experimental Electrochemistry for chemists*, Wiley-Interscience, 1974. Particularly recommended for its treatment of experimental conditions for all electro- analytical techniques.

(*b*) A J Bard and L R Faulkner, *Electrochemical Methods*, Wiley-Interscience, 1980. This represents a more advanced treatment of the subject and is at honours degree level and beyond.

(*c*) D R Crow, *Principles and Applications of Electrochemistry*, Chapmen and Hall 1979. An introductory text on electrochemistry that covers all the principles required.

(*d*) D B Hibbert and A M James, *Dictionary of Electrochemistry*, Macmillan, 1984.

3. GENERAL POLAROGRAPHY TEXTBOOKS

(*a*) D R Crow and J V Westwood, *Polarography*, Methuen, 1968. An excellent simple introductory text, good to start with.

(*b*) J Heyrovsky and J Kuta, *Principles of Polarography*, Interscience, 1965. Rather old but still a valuable general text on classical dc polarography.

(*c*) L Meites, *Polarographic Techniques*, Wiley-Interscienced, 1965. Although rather old, is particularly useful for its treatment of dc polarography as a practical subject and for the data listed.

(*d*) A M Bond, *Modern Polarographic Methods in Analytical Chemistry*, Marcel Dekker, 1980. A good modern general text with good coverage of modern techniques.

4. APPLICATIONS

(*a*) W F Smyth (Ed), *Polarography of Molecules of Biological Significance*, Academic Press, 1979. A good account of a range of practical applications.

(*b*) P Zuman, Topics in Organic Polarography, Plenum Press, 1970. One of a range of good books from this author, this one more practical than some.

(*c*) C L Perrin, 'Mechanisms of Organic Polarography' in S G Cohen, A Streitwieser and R W Taft (Eds), *Progress in Physical Organic Chemistry*, Vol.3, Wiley, 1965. A useful review of the range of electroactive organic compounds.

1. Classical DC Polarography

Overview

In this part the emphasis is on introducing you to the physico-chemical principles that underlie all of the polarographic methods of analysis. There are now several forms of polarography, the more important of which are dealt with in later parts of this unit.

The earliest form of polarography was dc polarography and since all the original theory was developed in terms of this technique, we will use it in order to establish principles.

In addition to these theoretical principles we shall discuss typical solution conditions, cell design and electrical circuitry most of which are factors common to all forms of polarography.

Central to the whole treatment will be the dropping mercury electrode (DME) and you will be given details of the physical and electrical properties of this important electrode.

Finally applications are considered and dc polarography is assessed as an analytical method. A discussion of the limitations of the method leads naturally to the content of the rest of the unit.

1.1.   INTRODUCTION

Classical dc polarography is now rarely used for analysis in modern industry, the method is in the main confined to teaching laboratories in academic institutions. Why then do we choose to begin this unit with a discussion of the dc method?

We will see later (3.0 and 4.0) that there are important modern electro-analytical methods based on the principles of dc polarography. We study dc polarography in order to learn these principles as a basis for the understanding of more recent advanced methods.

What is polarography?

In an electrolysis cell the current produced in the cell is a measure of the amount of chemical change occurring at the electrodes. This is a well known principle embodied in Faraday's Laws. In order for a reaction to occur at an electrode the electrode potential must exceed a certain critical value. There are a variety of electro-analytical techniques which are designed to measure cell current as a function of electrode potential. These techniques all belong to the branch of the electro-analytical techniques family known as voltammetry at finite current. If the electrode potential is changed in a linear mode then we have linear sweep voltammetry (LSV). Linear sweep voltammetry may be carried out using a variety of electrode materials such as gold, platinum, carbon and also mercury. If, and only if, mercury in the form of a dropping mercury electrode (DME) is used as the working electrode, we have polarography. The working electrode (WE) in a cell is the electrode where the reaction of interest is occurring, eg in the analysis for Cu(II) ions it is the electrode where the reaction

$$Cu^{2+} + 2e \rightarrow Cu$$

occurs. The mercury electrode in polarography takes a special form consisting of small droplets of mercury generated at the lower end of a glass capillary tube. This is termed a dropping mercury electrode (DME).

It might be useful here to look at the family of electro-analytical methods. These may be classified as follows.

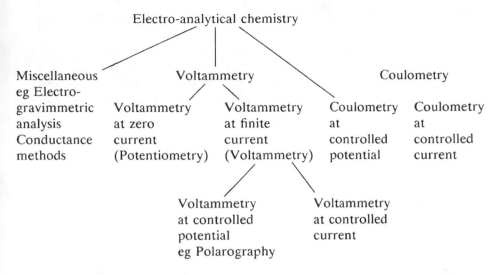

Polarography is then an example of voltammetry at controlled potential in which the working electrode consists of a dropping mercury electrode and the potential of the working electrode is changed in a linear mode. We shall not be concerned in this unit with any other parts of the above family of methods. Students who wish to obtain a broad overview of the whole family of methods should study ACOL: *Principles of Electro-analytical Methods.*

Why did dc polarography rapidly disappear from analytical laboratories in the mid-1950's? The main application of dc polarography in analysis was to heavy metal cation analysis and the detection limit here is about 10^{-4} mol dm^{-3} (*ca* 6 ppm for copper). In the mid 1950's atomic absorption spectroscopy (AAS) became available and was routinely used to analyse for heavy metal cations at <1 ppm. Atomic absorption spectroscopy, although more expensive in initial costs, is an easier technique to use than polarography. Atomic absorption spectroscopy became rapidly established as the method of choice and has remained so until the present day. It is only in recent years that advanced forms of polarography using pulse techniques have begun to become competitive with AAS (see Section 3.0 and 4.0).

During these years dc polarography has continued to play a minor role as a method for the analysis of organic materials and also to provide evidence for the mechanism of electrode reactions, particularly those involving organic species.

It remains in this introduction to give credit to the man who discovered, reported and established the technique. This was the Czechoslovakian chemist Jaroslav Heyrovsky in 1922. Czechoslavakia remains to this day a centre of strong activity in electro-analytical chemistry.

SAQ 1.1a	Explain what feature of polarography sets it apart from other voltammetric techniques.

SUMMARY AND OBJECTIVES

Summary

dc polarography has been related to other electro-analytical methods in general and to voltammetry at controlled potential in particular.

It has been established that by definition a dropping mercury electrode is necessary for the name polarography to be used.

Although dc polarography is no longer an important analytical technique, the theory is relevant to more advanced polarographic methods which are of importance.

Objectives

You should now be able to:

- state the relative importance of dc polarography as an analytical method;

- identify the place of dc polarography in the family of electro-analytical methods;

- state the detection limit for metal cation analysis using dc polarography;

- state the name of the discoverer of the dc polarography method of analysis.

1.2. CIRCUITRY AND CELL DESIGN

At the time that dc polarography went out of fashion in the 1950's a two-electrode circuit was used. Even if it had continued as a competitive technique three-electrode circuitry would have taken over by the mid-1960's. We will begin with a criticism of the two-electrode circuit and explain the benefits of the use of three-electrodes.

The essential features of good cell design will then be presented. It should be emphasised that the change to three-electrode circuitry and good cell design have no effect whatsoever on the inherent weakness of the dc technique, ie the high limit of detection.

1.2.1.　The Problems Associated with Two-electrode Circuitry

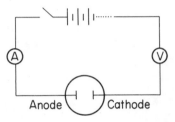

Fig. 1.2a.　*Two-electrode electrolysis cell*

Consider the circuit (Fig. 1.2a). In all electrolysis-based analytical methods we are interested in the reaction at one electrode only. This electrode is termed the *working electrode* (WE) and it may be either an anode or a cathode depending on the nature of the analytical reaction. For example if the reduction of copper(II) ions,

$$Cu^{2+} + 2e \rightarrow Cu,$$

is the analytical reaction then the WE is a cathode. In a two-electrode circuit, the electrode other than the WE, is termed the *secondary electrode* (SE) (alternative names are auxiliary electrode and counter electrode). There is a tendency to ignore the chemistry occurring at the SE. This can sometimes cause problems in other electro-analytical methods but rarely in polarography. The voltmeter reading in the circuit of Fig. 1.2a is a measure of the difference between the electrode potentials of the two electrodes together with the ohmic drop (IR) due to the resistance (R) of the electrolyte between the electrodes

$$V = E(\text{anode}) - E(\text{cathode}) + IR \qquad (1.2a)$$

where V is the voltmeter reading, ie the cell voltage.

∏   If the cell current in a two-electrode cell is 10 μA and the cell resistance is 20 Ω, what are the potentials of the anode and cathode when the voltmeter reads 2 V?

There is no answer to this question and that in itself makes an essential point. If you prefer, the answer is that it is impossible to calculate $E(\text{anode})$ and $E(\text{cathode})$.

You may calculate

$$IR = 10 \times 10^{-6} \times 20 = 2 \times 10^{-4} \text{ V.}$$

$\therefore \quad V = E(\text{anode}) - E(\text{cathode}) + IR$

$\qquad = E(\text{anode}) - E(\text{cathode}) + 2 \times 10^{-4}$

$\qquad = 2 \text{ V}$

$\therefore \quad E(\text{anode}) - E(\text{cathode}) \approx 2 \text{ V}$

However, you cannot now separate the contributions of $E(\text{anode})$ and $E(\text{cathode})$.

In polarography as we shall see later, the solutions used are usually aqueous solutions of strong electrolytes and in these cases the ohmic drop is negligible. If we ignore this factor then:

$$V = E(\text{anode}) - E(\text{cathode}) \qquad (1.2b)$$

but the problem of separating the anodic and cathodic contributions remains. If one of the electrodes has a potential which remains effectively constant as the cell voltage and current vary, then it may be regarded as a pseudo-reference electrode.

We have

$$V = |E_{WE} - E_{SE}| \qquad (1.2c)$$

The modulus sign allows for the WE to be either an anode or a cathode. If E_{SE} remains effectively constant (electrode unpolarised) we may state a value for E_{WE} with respect to the potential of the SE as a reference.

∏   If the cell voltage is 2 V and the ohmic drop is negligible, state the potential of a cathode WE with respect to the SE whose potential may be assumed constant.

The answer is $E_{WE} = -2$ V with respect to the potential of the secondary electrode.

$$V = E(\text{anode}) - E(\text{cathode}) + IR$$

$$= E_{SE} - E_{WE} + 0$$

$$\therefore \quad E_{WE} - E_{SE} = -2 \text{ V}$$

Similarly if the cell voltage changes from V_1 to V_2 and the ohmic drop is negligible and $E(\text{anode})$ (SE) is constant then:

$$\Delta V = V_2 - V_1 = \left| (E_{2,WE} - E_{1,WE}) - (E_{2,SE} - E_{1,SE}) \right|$$

$$= \left| \Delta E_{WE} - 0 \right|$$

$$= \left| \Delta E_{WE} \right| \tag{1.2d}$$

Thus we have an important result that a change in the measured cell voltage reflects a change in E_{WE}.

Two-electrode circuitry prevailed into the late 1950's and the integrity of the method was dependent upon the potential of the secondary electrode remaining constant during the analysis. If the secondary electrode has a low resistance and a large area and the electrode reaction is such that the potential is insensitive to small concentration changes then the necessary condition of constant potential is met reasonably well. One such electrode is a pool of mercury in contact with an aqueous solution containing chloride ion. This was the electrode of choice for SE in dc polarography. However the uncertainties described above make any two-electrode system unsatisfactory.

If non-aqueous solvents are used then one can sometimes no longer ignore the ohmic drop and this quantity (IR) will vary as the cell voltage and cell current vary. In this situation even with E_{SE} a constant E_{WE} is unknown.

∏ Calculate the resistance of a 2 cm length of 0.1 mol dm^{-3} aqueous KCl between two electrodes, each of area 2 × 10^{-5} m^2. The conductivity κ of 0.1 mol dm^{-3} aqueous KCl is 1.29 S m^{-1}

The answer is $R = 775 \; \Omega$.

G = conductance κ/J,

where κ is the conductivity, J the cell constant, L/A where L is the path length and A the electrode area.

∴ $G = 1.29 \times 2 \times 10^{-5}/0.02 = 1.29 \times 10^{-3}$ S

∴ $R = 1/G = 775 \; \Omega$

For a typical dc polarography experiment the cell current would be 5–10 μA. This would give an ohmic drop of 4–8 mV.

It would be an improvement in technique if a circuit could be devised such that the potential of the working electrode (E_{WE}) could be measured and controlled in an unambiguous matter.

1.2.2. Three-electrode Circuitry

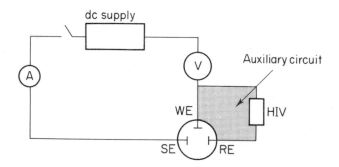

Fig. 1.2b.   *Three-electrode electrolysis cell (WE, working elec-trode; SE, secondary electrode; RE, reference electrode; HIV, high impedance voltmeter)*

The problems described in the previous section are largely over-come by the introduction of a third electrode into the circuit (Fig. 1.2b). An auxiliary circuit, often termed a potentiometer circuit, has been added to the previous two-electrode circuit. The cell current still passes between the WE and the SE in the primary circuit and the cell voltage (V) is still developed between these electrodes. We still have:

$$V = \left| E_{WE} - E_{SE} \right| + IR \qquad (1.2e)$$

However Eq. 1.2e is now redundant since we are interested only in E_{WE} and in these circuits the secondary electrode is typically a small platinum electrode with a potential which is not constant. In many modern instruments the cell voltage is not displayed.

Why are the Eq. 1.2e and the cell voltage redundant?

Consider the auxiliary circuit comprising the WE, the reference electrode (RE) and a high impedance voltmeter (HIV). If P is the potential difference measured by the HIV then:

$$P = E_{WE} - E_{RE} + IR' \qquad (1.2f)$$

where R' is the resistance in the electrolyte path between the WE and RE. This is a null current potentiometer circuit and not part of the primary current carrying circuit of the cell, an essential point to remember. Provided $IR' \rightarrow 0$, for the reasons given previously:

$$P = E_{WE} - E_{RE} \qquad (1.2g)$$

For a reference electrode we choose one of the well-established reference electrodes used in potentiometry, eg saturated calomel electrode (SCE), silver–silver chloride electrode. Such electrodes have a stable potential and a small temperature dependence of potential. The saturated calomel electrode is almost always the reference electrode used for aqueous solution dc polarography.

∏   Why did we not use a SCE as an electrode of fixed potential in a two-electrode circuit?

A SCE is a reversible electrode, small in size, and designed to give a fixed potential under null current conditions. If a finite current passes through such an electrode polarisation occurs and the potential changes. The electrode returns to its original potential when the current ceases to flow provided the current has flowed for only a short time.

We quote the values of the E_{WE} with respect to the potential of the reference electrode. Thus substituting E_{SCE} for E_{RE} in Eq. 1.2g gives:

$$P = E_{WE} - E_{SCE}$$

∴

$$P = E_{WE}(SCE) \qquad (1.2h)$$

If the measured voltage on the high impedance voltmeter is -0.600 V then $E_{WE} = -0.600$ V (SCE). Where does the minus sign come from? The voltmeter will either display both positive and negative values or the operator has to note the polarity of the connections made. Values of E_{WE} obtained with reference electrodes other than the SCE are usually converted to the SCE scale before reporting the results.

Reaction		$E^{\ominus}/V(NHE)$
$Co^{3+} + e$	$\rightarrow Co^{2+}$	1.821
$Ce^{4+} + e$	$\rightarrow Ce^{3+}$	1.612
$Cl_2 + 2e$	$\rightarrow 2\,Cl^-$	1.358
$O_2 + 4\,H^+ + 4e$	$\rightarrow 2\,H_2O$	1.229
$Cu^{2+} + 2\,CN^- + e$	$\rightarrow Cu(CN)_2^-$	1.120
$Br_2 + 2e$	$\rightarrow 2\,Br^-$	1.087
$2\,Hg^{2+} + 2e$	$\rightarrow Hg_2^{2+}$	0.905
$Ag^+ + e$	$\rightarrow Ag$	0.800
Hg_2^{2+}	$\rightarrow 2\,Hg$	0.789
$Fe^{3+} + e$	$\rightarrow Fe^{2+}$	0.771
$O_2 + 2\,H^+ + 2e$	$\rightarrow H_2O_2$	0.682
$Hg_2SO_4 + 2e$	$\rightarrow 2\,Hg + SO_4^{2-}$	0.616
$I_2 + 2e$	$\rightarrow 2\,I^-$	0.536
$Cu^+ + e$	$\rightarrow Cu$	0.521
$O_2 + 2\,H_2O + 4e$	$\rightarrow 4\,OH^-$	0.401
$Cu^{2+} + 2e$	$\rightarrow Cu(Hg)$	0.345
$Cu^{2+} + 2e$	$\rightarrow Cu$	0.340
$Hg_2Cl_2 + 2e$	$\rightarrow 2\,Hg + 2\,Cl^-$	0.268
$Ag\,Cl + e$	$\rightarrow Ag + Cl^-$	0.222
$Cu^{2+} + e$	$\rightarrow Cu^+$	0.158
$Sn^{4+} + 2e$	$\rightarrow Sn^{2+}$	0.150
$Hg_2Br_2 + 2e$	$\rightarrow 2\,Hg + 2\,Br^-$	0.141
$AgBr + e$	$\rightarrow Ag + Br^-$	0.071
$2\,H^+ + 2e$	$\rightarrow H_2$	0.000
$Pb^{2+} + 2e$	$\rightarrow Pb(Hg)$	-0.121
$Pb^{2+} + 2e$	$\rightarrow Pb$	-0.126
$Sn^{2+} + 2e$	$\rightarrow Sn$	-0.136
$Tl^+ + e$	$\rightarrow Tl$	-0.336
$Cd^{2+} + 2e$	$\rightarrow Cd(Hg)$	-0.352
$Cd^2 + 2e$	$\rightarrow Cd$	-0.403
$Fe^2 + 2e$	$\rightarrow Fe$	-0.409
$Zn^2 + 2e$	$\rightarrow Zn$	-0.763
$2\,H_2O + 2e$	$\rightarrow H_2 + 2\,OH^-$	-0.828
$K^+ + e$	$\rightarrow K$	-2.924

Fig. 1.2c.   *Standard electrode potentials at 25 °C*

NB 1. $Hg_2Cl_2 + 2e \rightarrow 2\,Hg + 2\,Cl^-$ (saturated)

is the saturated calomel electrode (SCE) for which $E = 0.244$ V (NHE).

2. $AgCl + e \rightarrow Ag + Cl^-$ (saturated)

is the saturated silver chloride electrode and has $E = 0.199$ V (NHE).

∏   Convert the following results to the SCE scale, all at 25 °C.

You will need to consult the table of standard electrode potentials (Fig. 1.2c).

(a) $E_{WE} = +0.012$ V (saturated aq. Ag, AgCl/Cl$^-$)

(b) $E_{WE} = -0.230$ V (NHE)

Answers are (a) -0.033 V (SCE), (b) -0.474 V (SCE).

$E(Ag,AgCl,Cl^-, sat) = +0.199$ V (NHE) and

$E_{SCE} = +0.244$ V (NHE) from Fig. 1.2c.

∴ (a) $E_{WE} = +0.012 + 0.199 - 0.244 = -0.033$ V (SCE)

(b) $E_{WE} = -0.230 + 0 - 0.244 = -0.474$ V (SCE)

We now have a reliable measure of E_{WE}. Also,

$$\Delta P = \Delta E_{WE} \qquad (1.2i)$$

So we can monitor changes in E_{WE}. It is as well to remind you that an uncertainty arises if solvent systems of high resistance are used which makes the ohmic drop (IR') no longer negligible. This is usually due to the use of non-aqueous solvents and these can also cause the reference electrode to become unstable.

The next advance in technique was to be able to automatically select, change and monitor the E_{WE} in a controlled manner. This advance had to wait until an electronic control device called a potentiostat became available.

1.2.3. Potentiostatic Control

It has already been explained that dc polarography is not used to any great extent nowadays but when it is used the modern commercial apparatus incorporates a potentiostat which effectively controls the instrument. In order to fully understand the mode of operation of a potentiostat some knowledge of electronics is required. This knowledge must extend at least to an understanding of operational amplifiers. Those of you with this knowledge should refer to the textbooks listed in the reference section where you will find circuit diagrams and further explanation. Fortunately, it is not necessary to have this depth of understanding to appreciate the function of a potentiostat. This is to control the potential of the WE with respect to E_{RE}. The potentiostat enables one to hold this potential constant or to vary the potential in a controlled manner to a pre-selected pattern. To do this a three-electrode circuit is necessary (Fig. 1.2d).

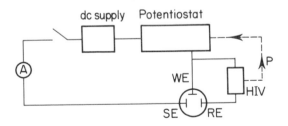

Fig. 1.2d. *Potentiostatic control*

$$P = \text{potential difference} = |\,E_{WE} - E_{RE}\,|$$

The required E_{WE} (SCE) is fed into the potentiostat control and when the circuit is completed the measured HIV reading is:

$P = E_{WE}$ (SCE) (measured),

and this value is fed instantly to the potentiostat. The potentiostat compares the measured value of E_{WE} with the required value and the difference is an error signal. If the error signal is zero the system is under potentiostatic control. If the error signal is finite the potentiostat causes a change in the dc power supply to the cell to occur in such a direction as to decrease the error signal. The response is very rapid and potentiostatic control is normally achieved within microseconds.

An extension of this is to have available the capability to scan a range of potential in a pre-determined manner. The required initial E_{WE} (SCE) is fed into the potentiostat control together with a chosen final E_{WE} (SCE) and the scan pattern. For example:

Initial E_{WE} (SCE) = 0 V

Final E_{WE} (SCE) = −2.0 V

Linear potential scan = 5 mV s⁻¹

In this case when the circuit is completed the initial E_{WE} (SCE) is rapidly established as described above. The scan commences and the potentiostat continuously monitors and adjusts E_{WE} (SCE) to conform to the selected pattern.

∏   For the linear scan pattern, given in the above example, calculate the reading on the HIV that will cause a zero error signal 2 min after the scan commences.

The answer is −0.60 V.

A zero error signal occurs when,

E_{WE} (SCE) measured = E_{WE} (SCE) (required).

The required E_{WE} (SCE) after 2 min is

$$0 - 2 \times 60 \times 5 \times 10^{-3}$$

$$= -0.60 \text{ V}$$

$$\therefore \quad P = E_{WE} \text{ (SCE) (measured)} = -0.60 \text{ V}$$

We will see later that a linear potential scan at about 2–10 mV s^{-1} is always used in dc polarography. Potentiostatically controlled three-electrode circuits have been used for dc polarography since the early 1960's. You should note that similar potentiostatically controlled circuits are widespread in modern electro-analytical chemistry.

1.2.4. Cell Design

Basically we require a three electrode cell, Fig. 1.2e.

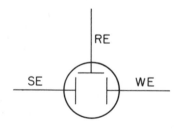

Fig. 1.2e. *Schematic design of three-electrode cell*

Usually there are other requirements, commonly provision is made for some or all of the following:

(*a*) thermostatting,
(*b*) emptying the cell *in situ*,
(*c*) adding chemicals,
(*f*) purging with gas,
(*g*) gas venting.

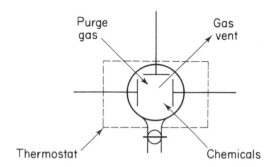

Fig. 1.2f. *Schematic diagram of three-electrode cell with additional requirements*

A well designed cell will provide these features together with minimisation of cell volume. A suitable cell for dc polarography is shown in Fig. 1.2g. Note the pear shape used to minimise the volume. We may pick out two features for a more detailed treatment.

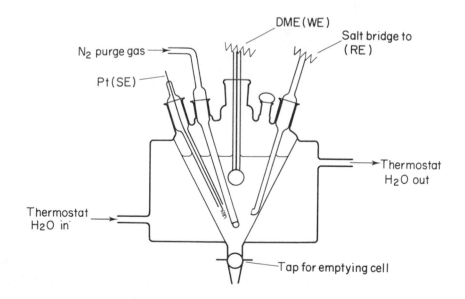

Fig. 1.2g. *Typical dc polarography cell*

(*i*) Device for gas purging

The purging gas, usually nitrogen, may be introduced via a two-way
tap in a device depicted in Fig. 1.2h. This design is given as an
example. In one position of the tap, gas flows down the tube and
through a frit into the base of the analyte solution. In the second
position the gas enters into the neck of the cell and blankets the surface of the analyte solution thus preventing re-entry of air into the
solution. Tank nitrogen (white spot) contains a few ppm of oxygen
and if it is necessary to further reduce the oxygen level in the solution the nitrogen supply must be scrubbed of oxygen. One method
of doing this is to pass nitrogen through a vanadium(V) chloride
solution (purple). Alternatively commercial cartridges filled with a
scrubbing formulation based on chromium(III)oxide are available
as disposable inserts into the gas stream.

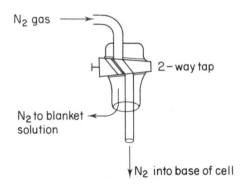

As drawn N_2 flows into base of cell

Fig. 1.2h.   *Gas purging and blanketing device*

(*ii*) Reference electrode bridge

In some circumstances the reference electrode, eg SCE may be introduced direct into the cell. The only physical separation of the
electrode inner solution from the analyte solution is a small frit or
a piece of glass wool. Seepage from the electrode can occur and this
may create interference with the analytical reaction. To overcome

this, and also as part of the usual design practice, the electrode is placed in a separate compartment, and a salt bridge is formed to the analyte solution (Fig. 1.2i).

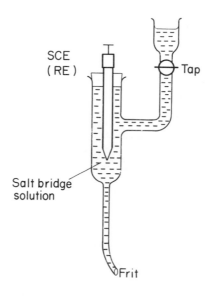

Fig. 1.2i. *Reference electrode salt bridge*

The salt bridge solution is usually the supporting electrolyte solution used for the analysis, (1.4). Mention should be made of a special form of such a reference electrode system where the tip of the salt bridge is bent to a position close to the WE (Fig. 1.2j).

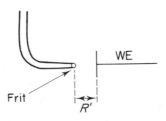

Fig. 1.2j. *Luggin probe*

This device is called a Luggin probe and its purpose is to minimise the ohmic drop (IR'). This may become important for non-aqueous solutions.

SAQ 1.2a

Sketch and label a circuit for a potentiostatically controlled three-electrode cell and explain how potentiostatic control is achieved and maintained. Why are three electrodes necessary? What problems arise when non-aqueous solutions are used?

SAQ 1.2b   Sketch the appearance of a cell suitable for dc polarography analysis and comment on the features.

SUMMARY AND OBJECTIVES

Summary

A criticism of two-electrode circuitry is presented based on the uncertainty of the value of E_{WE}. The use of three electrodes removes this uncertainty and produces an unambiguous value for E_{WE} (SCE). The use of a potentiostat to control the value of E_{WE} (SCE) is described. Finally the necessary components of a polarography cell are listed and considered from the point of view of good cell design.

Objectives

You should now be able to:

● criticise the use of two-electrode circuitry for dc polarography;

● draw a circuit diagram (fully labelled) for use with a three-electrode cell and explain the function of the auxiliary (potentiometer) circuit;

● define WE, SE, and RE;

● explain the function of a potentiostat and draw a circuit diagram (fully labelled) showing the use of a potentiostat in controlling the WE potential;

● summarise the essential features of a well designed polarography cell.

1.3.   THE DROPPING MERCURY ELECTRODE (DME)

The unique feature of polarographic methods that sets them apart from other voltammetric techniques is the use of a dropping mercury electrode as the working electrode. In this section we shall deal with the properties of mercury as an electrode material and the characteristics of the electrode in use.

1.3.1. Mercury as an Electrode Material

Mercury is very toxic and demands a neat and tidy working practice. Polarography had gained a bad reputation by the 1950's in academic institutions, due mainly to careless use of mercury and the growing awareness of health hazards. This in turn has contributed to a resistance to the introduction of newer electro-analytical methods using mercury even when they are competitive or better than established methods. This is a pity because there should be no hazard provided reasonable care is taken.

Mercury is a liquid under normal experimental conditions and thus presents a smooth homogeneous surface to the solution. It is a good electrical conductor and provides a surface for the required electrode reaction. A good electrode material provides a high voltage limit in anodic and cathodic directions and thus a wide voltage window for analysis.

What are voltage limits and the voltage window?

From a practical point of view the potential of the working electrode at which a measurable current begins to develop, as a consequence of some process other than the desired analyte reaction, defines the voltage limit for the system. The unwanted reaction will interfere with any required analyte reaction that occurs at a potential beyond this limit. There will be such a limit in the anodic and the cathodic directions. The difference between these limits defines the voltage window. It is possible to analyse for any species which can be oxidised or reduced at a potential inside the voltage window.

What determines the voltage window for mercury under the experimental conditions of dc polarography?

In the anodic direction the limit is set by the tendency of mercury to oxidise at potentials more positive than -0.1 V (SCE), the actual potential depending on the composition of the solution and never exceeding $+0.2$ V (SCE). In the cathodic direction the limit is set by the reduction of the most readily reduced species in the solution. In aqueous solutions this is one of three reactions:

$$H_3O^+ + e \rightarrow \frac{1}{2}H_2 + H_2O$$

$$H_2O + e \rightarrow \frac{1}{2}H_2 + OH^-$$

$$ox + ne \rightarrow red$$

Which of the first two reactions prevails depends upon the pH of the solution and the potential varies from about -0.8 V (SCE) to -2 V (SCE), the lower the pH the less negative the limit. It is worthy of comment why the potential for the reduction of the hydronium ion is as negative as -0.8 V (SCE). Using the Nernst equation you should be able to predict that this potential should be about -0.4 to -0.6 V (SCE) depending upon the hydronium ion concentration. This is the potential range experienced when a platinum electrode is used. On this electrode the reaction is very fast and the reversible potential is observed. On mercury the reaction is slow and this gives rise to an overpotential causing the observed potential to be much more negative. This is a distinct advantage for mercury since it extends the cathodic voltage limit. Overpotential is mentioned again (1.5.1) but it is assumed that the concept is familiar to you. If not read the relevant sections in ACOL: *Principles of Electroanalytical Methods* or in Crow (1974). The reaction:

$$ox + ne \rightarrow red$$

is the reduction of any reducible species present in the system. We will see later (1.4) that in dc polarography a high concentration of potassium chloride is often added to the solution and for potassium ions the voltage limit for reduction is about -2 V (SCE).

We see then that the voltage window for mercury in aqueous solutions could be, at best, $+0.2$ V (SCE) to -2 V (SCE).

A further important factor must be taken into account when discussing the available voltage window. All solvents dissolve oxygen to some extent and protic solvents, eg H_2O, dissolve appreciable amounts (about 10^{-3} mol dm^{-3} or 30 ppm in water at 25 °C). Oxygen is electroactive and is reduced in aqueous solutions in two stages at potentials of about -0.5 V (SCE) and about -0.6 to -1.2 V (SCE), the latter value depending upon the acidity of the solution.

Acidic conditions:

$$O_2 + 2H_3O^+ + 2e \rightarrow H_2O_2 + 2H_2O$$

$$O_2 + 4H_3O^+ + 4e \rightarrow 6H_2O$$

Neutral or alkaline conditions:

$$O_2 + 2H_2O + 2e \rightarrow H_2O_2 + OH^-$$

$$O_2 + 2H_2O + 4e \rightarrow 4OH^-$$

These reactions give rise to two kinds of problem. The most important of these is that the signal generated by the reactions (observed current) masks the required signals and renders dc polarography useless in the range -0.5 V to at least -1.0 V (SCE). This is virtually the whole of the useful range of potential for the method. The second type of problem arises from the direct interference of the reaction products, ie H_2O_2 and/or OH^-, with the analyte reaction. An example would be a localised rise in pH due to the generation of hydroxyl ions at the cathode. It is thus very important to remove oxygen from the solution prior to analysis if a cathodic process is to be investigated. For methods see 1.2.4.

It is important to use clean mercury and there are three classes of impurity likely to be found in laboratory mercury: surface scum (mainly oxide), dissolved base metals and dissolved noble metals. The surface scum is removed by repeated filtration through a small perforation in a filter paper cone. Base metals, eg zinc, may be removed by drawing air through the mercury under 2 mol dm^{-3} aqueous nitric acid. The only method satisfactory for removal of noble metals is by distillation and it is recommended that this is carried out by a specialist mercury supplier. It should only be necessary to periodically carry out the first two operations for satisfactory recycling of mercury for normal laboratory use. Scum will impare flow through the capillary, base metal reduction peaks will interfere in the analysis and noble metals will cause the cathodic voltage limit to become less negative since they are better catalysts than mercury for the reduction of hydronium ions.

1.3.2. Physical Characteristics of the Electrode

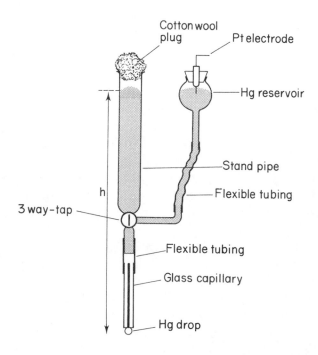

Fig. 1.3a. *Typical dropping mercury electrode*

Fig. 1.3a shows the design of a typical dropping mercury electrode. The height of the column of mercury above the position of the mercury drop (h) is adjusted in the range 20–100 cm. The capillary is usually 10–20 cm long with an internal diameter of about 50 μm. The mature mercury drop achieves a diameter of about 1 mm just prior to dropping. The mass of mercury produced per unit time (m/kg s^{-1}) may be calculated using the Poiseuille equation.

$$m = P\pi\rho r^4/8L\eta \qquad (1.3b)$$

where

$P/\text{kg m}^{-1}\text{ s}^{-2}$ is the pressure due to the height of the mercury column, $\rho/\text{kg m}^{-3}$ the density of mercury, r/m the radius of the capillary tube, $\eta/\text{kg m}^{-1}\text{ s}^{-1}$ the coefficient of viscosity of mercury and L/m the length of the capillary tube.

The value of the pressure P should be corrected for the effect of two factors:

— a small but significant effect of the surface tension at the Hg/H_2O interface which varies with the value of the electrode potential, about 1–2 cm Hg;

— a very minor back pressure due to the depth of immersion of the capillary into the aqueous solution; a few mm Hg.

A correction of 2 cm Hg is usually made and this is sufficiently accurate for most work. In analytical applications it is rarely necessary to know the value of h.

Now $P = \rho gh$,

$$m = \pi g\rho^2 r^4 h/8L\eta \tag{1.3c}$$

The significant point is that, $m \propto h$

The lifetime of each drop of mercury, the drop lifetime (t/s) is of great importance in the theory of polarography and is usually arranged to be in the range 0.5–5 s. The mass of each drop $= mt$, and this quantity remains fixed, ie as h is increased so m increases and hence t falls and *vice versa*.

$$t \propto m^{-1} \propto h^{-1} \tag{1.3d}$$

∏   The mass of 30 drops of mercury from a DME was found to be 0.203 g and the drop lifetime was found to be 3.86 s. Calculate the flow rate of mercury from the capillary in mg s^{-1}.

The answer is 1.75 mg s^{-1}.

This is a very straight forward calculation.

Each drop weighs $\dfrac{0.203}{30}$ g

$$= \dfrac{0.203}{30} \times 10^3 \text{ mg}$$

$\therefore$ flow rate $(m) = \dfrac{0.203 \times 10^3}{30 \times 3.86}$

$$= 1.75 \text{ mg s}^{-1}$$

∏   If the height of the mercury column in the above problem was 30 cm and this height is changed to 90 cm, what will be the new drop lifetime?

The answer is 1.29 s.

The flow rate $m \propto h$

$\therefore$ the new flow rate $= 1.75 \times 90/30$

$$= 5.25 \text{ mg s}^{-1}$$

mt = constant $= 1.75 \times 3.86$

$\therefore$ the new drop lifetime $= \dfrac{1.75 \times 3.86}{5.25}$

$$= 1.29 \text{ s}$$

ie $t \propto h^{-1}$

$\therefore$ $t = 3.86 \times \dfrac{30}{90} = 1.29 \text{ s}$

We have concentrated on explaining the control over drop lifetime that may be exercised by adjustment of the head of mercury. You will learn later that more advanced techniques depend on the absolute reproducibility of this droptime. This necessitates the introduction of a mechanical tapping device controlled by the potentiostat which can produce drops with lifetimes of say 0.5, 1 or 2 s with great precision. Since this facility is built into modern instrumentation it is now also routinely used for dc polarography.

1.3.3. Electrical Characteristics of the Electrode

We have seen (1.3.1) that the voltage window for the DME is usually from about $+0.2$ V (SCE) to -2 V (SCE) with small variations due to differing solution conditions. In addition to this very important property there are electrochemical consequences of the effect of changing potential on the interface formed between the mercury drop and the aqueous solution.

(*i*) The capacitive current

An electrode at a certain potential has an electrical charge at its surface. For example in Fig. 1.3b we illustrate the situation when the electrode has a positive charge (cathode).

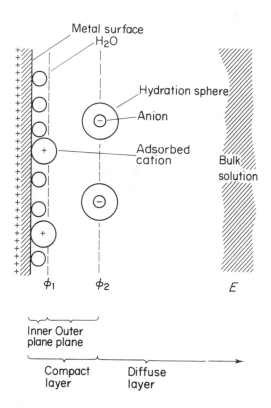

Fig. 1.3b. *The electrical double layer*

We see that there is a complex structure in the solution, changing as we move from the electrode surface to the bulk solution. There have been several theories over the years attempting to explain this structure and they all provide a model of the electrical double layer. Fig. 1.3b resembles the model proposed by Stern in 1924 and is sufficiently advanced for our purposes. We see that there are three zones reaching out from the electrode, each zone having a potential defined by ϕ_1 and ϕ_2 and finally E (the electrode potential, ie difference in potential between the electrode and the bulk solution).

Zone 1 is the inner plane consisting mainly of adsorbed cations and water molecules.

Zone 2 is the outer plane consisting mainly of solvated anions.

Together they form what is called the compact layer which usually has a thickness of about 5×10^{-10} m and varies only slightly with electrolyte concentration.

Zone 3 forms the diffuse layer which eventually merges into the bulk solution. The thickness of the diffuse layer is very dependent upon the electrolyte concentration becoming very thin (about 10^{-8} m) in 0.1 mol dm^{-3} solutions.

Because we have the model of parallel charged layers of different potential we have a direct analogy with a parallel plate condenser and it is therefore perhaps no surprise that a capacitance exists across the double layer.

Capacitance of   Capacitance of   Capacitance of
double layer (C_{DL}) = compact layer + diffuse layer
$\qquad\qquad\qquad\qquad$ (C_{cl}) $\qquad\qquad\qquad$ (C_{dl})

$$C_{DL} = dQ/dE, \qquad\qquad (1.3e)$$

where Q is the total charge in all layers ie the charge on the electrode, and E the electrode potential.

Fig. 1.3c shows how the double layer capacitance varies with the potential of the electrode.

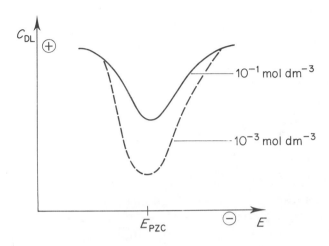

Fig. 1.3c. *Effect of potential on double layer capacitance*

We see that the detailed behaviour varies with the electrolyte concentration but there is always a minimum in the relationship which occurs at a potential characteristic of the electrode material/solvent interface. This is the potential of zero charge (E_{PZC}). The potential of zero charge is the potential at which the sign of the charge on the electrode changes. For the Hg/H_2O interface E_{PZC} is about -0.5 V (SCE). For the Hg/H_2O interface the double layer capacitance, C_{DL}, is about $20-25$ μF cm^{-2}.

If we define

$$E(PZC) = E - E_{PZC}$$

then

$$Q = C_{DL} A E(PZC) \qquad (1.3f)$$

We have taken C_{DL} to be expressed per unit area/F m^{-2} and A/m^2 is the area of the electrode.

Now $I = dQ/dt$ in general.

$\therefore$ From 1.3f,

$$I_{cp} = C_{DL} A \frac{\delta E(PZC)}{\delta t} + C_{DL} E(PZC) \frac{\delta A}{\delta t} + AE(PZC) \frac{\delta C_{DL}}{\delta t} \quad (1.3g)$$

charging of the double layer	effect of varying drop size	random effects during lifetime of drop eg adsorption of ions

Thus I_{cp}, the capacitive current contains three contributions. However under the conditions of dc polarography the last effect is usually absent. Also the rate of scan of potential (2–10 mV s^{-1}) is such that the drop is at approximately the same potential during its lifetime. Therefore during the lifetime of any one drop,

$$I_{cp} \approx C_{DL} E(PZC) \delta A / \delta t \quad (1.3h)$$

∏ Is it true that in dc polarography the drop is approximately at the same potential during its lifetime?

You should tackle this question by looking at the limiting cases. We have said that potential scan rates lie in the range 2–10 mV s^{-1} and that drop lifetimes are in the range 2–8 s. The worst limiting case is 10 mV s^{-1} scan and 8 s lifetime. Here the potential changes by 80 mV clearly not a constant potential. With 2 mV s^{-1} and 2 s we have only a 4 mV change.

This is effectively constant. So in practice the first term in Eq. 1.3g may not be negligible.

Treating the Hg drop as a sphere,

$$V = (4/3) \pi r^3 = mt/\rho$$

and $A = 4\pi r^2$.

Hence we may show that

$$A = (4\pi)^{1/3} \ 3^{2/3} \ \rho^{-2/3} \ m^{2/3} \ t^{2/3},$$

$$= k \ t^{2/3}$$

$\therefore \qquad \delta A / \delta t = (2/3) \ k \ t^{-1/3} = k' \ t^{-1/3}$

$\therefore \qquad$ From 1.3h $I_{cp} \approx k' \ C_{DL} E(PZC) \ t^{-1/3}$

This time dependence of the relationship is depicted in Fig. 1.3d and it is very important to note that the capacitive current decays very rapidly becoming negligible towards the end of the drop lifetime. This is the point to remember and you will find that we use this fact in later parts of the Unit (3.0).

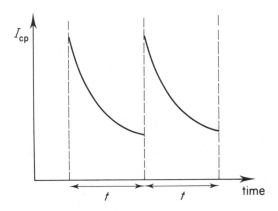

Fig.1.3d. *Capacitive current during lifetime of drop (t)*

The overall relationship of I_{cp} to E(SCE) in the range $+0.2$ V (SCE) to -2 V (SCE) is shown in Fig. 1.3e. The oscillations are caused by the decay of I_{cp} during the lifetime of each drop.

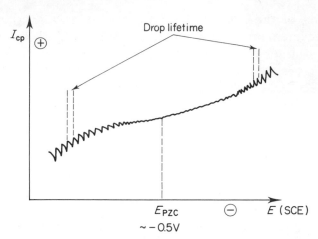

Fig. 1.3e. *Capacitive current as a function of electrode potential*

(*ii*) Current maxima

We shall see later that a typical dc polarography result takes the form of a wave, (Fig. 1.3f). The normal result is shown together with anomalous maxima superimposed. Both of these maxima are due to an enhancement in the rate of mass transport of the analyte to the electrode caused by convective movements at the Hg/H_2O interface.

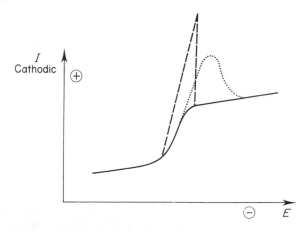

(———— normal result, wave; - - - - - - maximum of first kind; ········· maximum of second kind)

Fig. 1.3f. *Current maxima*

— Maxima of the first kind. These appear as sharp straight line enhancements in the current at the beginning of the wave. They can be larger than the wave itself and drop abruptly back to the wave maximum. They are due to convective streaming of the electrolyte past the drop surface caused by differences in surface tension at different points of the drop surface.

— Maxima of the second kind. These appear as relatively small rounded enhancements in the current on the wave plateau. They occur in solutions of high electrolyte concentration and are most common when a high flowrate of mercury is used. It is thought that the actual disturbance within the mercury drop generates the convective movement of the electrolyte next to the drop.

The most important point for analytical applications is to be able to remove these maxima. It has been known for a long time that small concentrations of surface active agents suppress these maxima. Examples of chemicals used in the early days of dc polarography are gelatine and methyl red. More recently the non-ionic detergent Triton X-100 (Rohm and Haas Co., Philadelpia, Penn) has become established as the material of choice. A concentration of <0.002% w/v is usually sufficient. This level of concentration should not be exceeded without checking for distortion of the polarographic waves produced.

SAQ 1.3a	Summarise the factors that determine the voltage window for dc polarography in aqueous solution.

SAQ 1.3b   Sketch and describe the construction of a typical
                DME and relate the quantities *m*, *h* and *t*.

SAQ 1.3c   Explain the origin of a capacitive current in dc
                polarography and state the time dependence of
                this quantity.

SAQ 1.3d	Sketch the appearance of current maxima in dc polarography and explain how these unwanted signals can be avoided.

SUMMARY AND OBJECTIVES

Summary

The properties of mercury as an electrode material are discussed and the concepts of voltage limits and voltage window introduced. The reasons for removing oxygen from the solution are given and with oxygen removed the voltage window established as about $+0.2$ V to -2 V (SCE). Methods of cleaning mercury and the reasons for cleaning are given.

The physical and electrical characteristics of the dropping mercury electrode (DME) are discussed. The physical characteristics are explainable using the Poiseuille equation and in particular the relationship between drop lifetime and height of the mercury column established.

The effect of potential of the behaviour of the Hg/H_2O interface is discussed in terms of the electrical double layer and the potential of zero charge. This leads to the concept of a capacitive current and an equation is derived relating this current to the electrode potential and also to the drop lifetime.

The phenomena of current maxima are illustrated; the origins explained and a simple method of suppressing the effect given.

Objectives

You should now be able to:

● explain the meaning of voltage limit and voltage window in dc polarography and discuss the factors which determine the value of the voltage window for mercury in aqueous solutions;

● explain why oxygen must be removed from solutions prior to analysis;

● sketch a typical DME;

● state the Poiseuille equation and use it to show the relationships between m, h and t;

● sketch a model of the electrical double layer and define the potential of zero charge;

● show how the analogy between the electrical double layer and a parallel plate condenser leads to the concept of a capacitive current, derive an equation relating $I_{cp}, C_{DL}, E(PZC), A, t$;

- explain the origins of current maxima of the first and second kind, sketch the appearance of such maxima and explain how to suppress them.

1.4. SOLUTION CONDITIONS

This section will provide you with the information that will allow you to select the solvent and the supporting electrolyte for a particular analysis. The nature and role of this supporting electrolyte will be explained. The effect of temperature is mentioned and together with the knowledge gained from Sections 1.2 and 1.3 you will be in a position to design a dc polarography experiment.

1.4.1. Solvents

The very nature of electro-analytical methods based on the generation of current between two electrodes at different potentials, implies the passage of ionic species through an electrically conducting solution. This in turn dictates that conditions exist in solution capable of sustaining stable ionic species. This, of necessity, requires that the solvent has some degree of polar character in order to be capable of solvating the ions.

∏   What is the physical property of a solvent that is usually used as a measure of polarity?

You should have answered the relative permittivity (or dielectric constant) (see Fig. 1.4a).

Highly polar, H_2O, $\epsilon_r = 78.5$ at 25 °C
Non polar, CCl_4, $\epsilon_r = 2.2$ at 25 °C

Water, which is an example of a protic solvent, is the most important solvent for dc polarography. Other protic solvents used are methanol and ethanol and sometimes binary mixtures of these with water. The use of solvents other than water is normally necessary for applications in organic chemistry where the solubility of the analytes in water becomes a problem. For the same reason some aprotic solvents are used, the most common being acetonitrile, CH_3CN, dimethyl sulphoxide $(CH_3)_2 SO$, (DMSO) and dimethylformamide $HCON(CH_3)_2$, (DMF). These aprotic solvents are not usually used in mixtures with each other but are often used in mixtures with protic solvents, in particular with water.

Solvent:	H_2O	DMSO	DMF	CH_3CN	CH_3OH	C_2H_5OH
ϵ_r:	78.4	46.7	36.7	36.1	32.7	24.6

Fig. 1.4a.   *Values of the relative permitivity of some solvents at 25 °C*

∏   Explain what you understand by a protic and an aprotic solvent.

Protic solvents exchange protons rapidly and are strong hydrogen donors. The hydrogen atoms in these solvents are bound to more electronegative atoms. Aprotic solvents have hydrogen bound to carbon and are very weakly acidic. They are poor hydrogen bond donors and exchange protons only very slowly.

Most of the reported analytical work on dc polarography has been done using aqueous solutions because in the past metallic cation analysis dominated the field. As interest in the analysis of organic materials has grown and with the continuing use of polarographic methods in organic mechanism studies, the use of non-aqueous solvents has increased.

You are reminded (1.2.1) that it is necessary to have an electrically conducting solution with a low electrical resistance. Even water, which has the lowest resistance of the solvents discussed above, is not sufficiently conducting without the addition of an electrolyte. We will see shortly (1.4.2) that in dc polarography it is a necessary feature that a high concentration of an electrolyte (other than the analyte) is always present in the solution.

We have also seen (1.3.1) that the presence of hydronium ions in solution is a very important factor in determining the cathodic voltage limit, and any solvent containing water will contain hydronium ions. Finally the relatively high solubility of oxygen in protic solvents presents problems.

1.4.2. Supporting Electrolytes

A supporting electrolyte is added to all solutions used for dc polarography. These supporting electrolytes are strong electrolytes (strong in water) and are often referred to as indifferent or inert. These terms should be familiar to you from other areas of chemistry, eg the use of ion-selective electrodes. The terms indifferent or inert are used because the added ions must not participate directly in the required electrode reaction.

The supporting electrolyte has two roles to play.

(*a*) The concentration of the supporting electrolyte regulates the electrical resistance of the cell. Even with water, the most polar solvent used, the resistance across a liquid path of say 2 cm would be quite high. With aprotic solvents the electrical resistance is so high as to make electrolysis impractical unless a supporting electrolyte is added.

∏   Calculate the resistance of a 2 cm length of a solution of 1.0 mol dm^{-3} $(C_2H_5)_4NBF_4$ in CH_3CN between two electrodes each of area 2×10^{-5} m^2. The conductivity of the solution is 5.55 S m^{-1}.

The answer is 180 Ω

$$G = \text{conductance} = \kappa/J = \frac{5.55 \times 2 \times 10^{-5}}{0.02}$$

$$= 5.55 \times 10^{-3} \text{ S}$$

$\therefore$ Resistance, $R = 1/G = 180 \ \Omega$

Compare this with the question in 1.2.1.

(b) The supporting electrolyte controls the migration of ions between the working and the secondary electrode. It is the mobility of an ion, u, that determines the fraction of the total migration current carried by an ion and the transport number of an ion (t) is a measure of that fraction. In the bulk electrolyte the migration of ions under the potential gradient is the only net transfer process. Near the electrode however, concentration gradients may occur, particularly in unstirred solutions. These gradients give rise to a diffusion current. It is of great importance in dc polarography to understand the factors determining the relative magnitudes of the migration current (I_m) and the diffusion current (I_d).

Consider a worked example.

An aqueous solution which contains $Cu(NO_3)_2$, concentration 10^{-3} mol dm^{-3}, is electrolysed between a Pt anode and a Hg cathode at 25 °C. What are the relative magnitudes of I_m and I_d for the Cu(II) ion at the cathode?

Mobility data at 25 °C:

$$u(Cu^{2+}) = 6.2 \times 10^{-8} \text{ m}^2 \text{ V}^{-1} \text{ s}^{-1}$$

$$u(NO_3^-) = 7.4 \times 10^{-8} \text{ m}^2 \text{ V}^{-1} \text{ s}^{-1}$$

The total current at the electrode is $I = I_m + I_d$.

Let an arbitary number of Faradays of electricity pass, say 10 Faradays (10 mol of electrons), ie $I \propto 10$ units.

$$t(i) = \frac{z(i)u(i)c(i)}{\sum_j z(j)u(j)c(j)}$$

∴ $t(Cu^{2+}) =$

$$\frac{(2 \times 6.2 \times 10^{-8} \times 10^{-3})}{(2 \times 6.2 \times 10^{-8} \times 10^{-3}) + (1 \times 7.4 \times 10^{-8} \times 2 \times 10^{-3})}$$

$$= 0.46$$

∴ A fraction 0.46 of the total cathodic current will be due to migration of ions to the electrode,

ie, $I_m \propto 4.6$ units

∴ $I_d \propto 5.4$ units

Thus $I_m / I_d = 0.85$ for Cu^{2+} at the cathode.

A 10^{-3} mol dm^{-3} $Cu(NO_3)_2$ solution is typical of a solution of metallic cations that can readily be analysed using dc polarography. What is the effect of adding a strong electrolyte to this solution on the I_m / I_d ratio? Try the following problem using the above worked example as a model.

∏ An aqueous solution containing 10^{-3} mol dm^{-3} $Cu(NO_3)_2$ is electrolysed between a Pt anode and a Hg cathode at 25 °C. If KCl is added as a supporting electrolyte at a concentration of 0.1 mol dm^{-3} what are the relative magnitudes of I_m and I_d for the Cu(II) ion at the cathode?

Additional mobility data at 25 °C:

$u(K^+) = 7.6 \times 10^{-8}$ m^2 V^{-1} s^{-1},

$u(Cl^-) = 7.9 \times 10^{-8}$ m^2 V^{-1} s^{-1}.

This is a more complicated problem than the worked example but the principle is identical.

Answer is $I_m / I_d = 0.008$ ie $> 99.2\%$ diffusion.

$$t(Cu^{2+}) = \frac{2 \times 6.2 \times 10^{-8} \times 10^{-3}}{\text{Denominator}}$$

Denominator =

$(2 \times 6.2 \times 10^{-8} \times 10^{-3}) + (1 \times 7.6 \times 10^{-8} \times 10^{-1})$

$+ (1 \times 7.4 \times 10^{-8} \times 2 \times 10^{-3}) + (1 \times 7.9 \times 10^{-8} \times 10^{-1})$

$$= 15.2 \times 10^{-9}$$

$$\therefore \quad t(Cu^{2+}) = \frac{12.4 \times 10^{-11}}{15.2 \times 10^{-9}} = 0.008$$

As in the worked example if $I \propto 10$ units

$$\therefore \quad I_m \propto 0.08 \text{ units}$$

$$I_d \propto 9.92 \text{ units}$$

Thus $I_m/I_d = 0.008$

The cathodic process here is clearly diffusion controlled.

We see that the addition of a supporting electrolyte at a concentration greatly in excess of that of the analyte causes the ionic migration of the analyte ion to be reduced to a negligible value. Usually the supporting electrolyte is added in at least ten-fold excess and preferably 100-fold excess with the limit often set by the solubility. Under these conditions the predominant mechanism for transport of the analyte ion to the electrode in an unstirred solution is diffusion. It is an essential feature of dc polarography that diffusion control exists in the solution.

Some common supporting electrolytes used in aqueous or near aqueous solutions are: KCl, $LiClO_4$, HCl, $HClO_4$, KOH and NaOH together with buffer solutions based on weak organic acids and phosphates. The common supporting electrolytes used primarily in aprotic and aqueous aprotic solvents are: tetraalkylammonium salts, $R_4N^+X^-$, where eg $R = CH_3$, C_2H_5, $t\text{-}C_4H_9$; $X^- = ClO_4^-$, BF_4^-, Cl, Br^-, I^-.

It is very important for you to realise that the cathodic voltage limit is often set by the potential at which the supporting electrolyte cation is reduced. Typically for salts of the type KCl used in aqueous solutions, the limit is about -1.5 to -2 V (SCE). For the tetraalkylammonium salts in non-aqueous solvents the limit is about -2.5 to -3.0 V (SCE).

Finally a note of caution on the use of supporting electrolytes when you are working at trace or lower levels of analyte concentration. When AnalaR grade supporting electrolytes are used at a concentration level of 0.1 mol dm^{-3} the concentrations of impurities added inadvertently, eg heavy metals, is significant at the ppb level. This is not a problem with dc polarography but is in the more advanced forms of polarography discussed in later parts of this unit.

∏ AnalaR grade KCl is usually quoted by the manufacturer as having <5 ppm Pb as an impurity. What is the concentration of Pb (in ppb) introduced into solution when 0.1 mol dm^{-3} aqueous KCl is used as a supporting electrolyte?

$A_r(K) = 39.1$, $A_r(Cl) = 35.5$

The answer is 3.7×10^4 ppb or 0.037 ppm.

Assume that the manufacturer's statement means that lead is present as the soluble Pb(II) ion at the level of 5 ppm in the solid KCl.

A 0.1 mol dm^{-3} solution of KCl contains 7.46 g KCl in 1 dm^3.

$\therefore$ Mass of Pb(II) present in 1 dm^3 = $\dfrac{7.46 \times 5}{10^6}$ g

$\therefore$ Mass of Pb(II) in 10^6 cm^3 = $\dfrac{7.46 \times 5 \times 10^3}{10^6}$

$= 0.0373$ g

This assumes that the density of the solution is 1 g cm^{-3}

Therefore the concentration of lead is

0.037 ppm or 3.7×10^4 ppb.

1.4.3. Effect of Temperature

The main source of error if the cell temperature is not controlled is in the change in the value of the diffusion coefficient of the analyte ion. In general diffusion coefficients (D) increase about 1–2% for every degree rise in temperature. We will see (1.5) that this will cause the signal which is a measure of the analyte concentration to change also, in the same sense, and by the same amount. For accurate work the cell should be thermostatted (Fig. 1.2g) and the temperature regulated to $\pm$ 0.2 °C. For much routine work no thermostatting is used and in this case care should be taken to protect the cell from wide fluctuations in the ambient temperature.

1.4.4. Concluding Remarks

You should now be able to select a suitable solvent/supporting electrolyte system for your purpose. The limitations on choice are set by the solubility of the analyte and of the supporting electrolyte, the requirement for a low electrical resistance, and the necessity to have a voltage window available for the required analyte reaction. This latter limitation usually amounts to ensuring that an adequate cathodic voltage limit is available.

∏   Why is this latter point true?

> It is because in dc polarography most of the analyses are carried out with the required reaction a reduction process. Only those oxidations which can occur at potentials more negative than the anodic voltage limit for Hg, about 0.1 V (SCE) are accessible to dc polarography.

Together with the matters discussed in sections 1.2 and 1.3 you are now in a position to list the features of a typical dc polarography experiment.

∏   What are the typical features of a dc polarography experiment? List them without attempting any explanation.

> — Potentiostatically controlled three-electrode circuit.

— DME (WE); Pt (SE); SCE (RE).

— Cell of minimum volume with facilities for: purging of O_2 with N_2, and thermostatting.

— Suitable solvent.

— Suitable supporting electrolyte.

— Addition of a maximum suppressor.

Although you are now able to design a system and you have a wide choice of solvents and supporting electrolytes, take care not to over design. For example it is only necessary to provide a cathodic voltage limit just sufficiently negative to accommodate the analyte reaction. The system $H_2O/0.1$ mol dm^{-3} KCl/Pt (anode, SE)/DME (cathode, WE) is a well tried system for heavy metal cation analysis and rather like an old shoe this system would only be reluctantly discarded by analysts for this application.

SAQ 1.4a   Select a suitable solvent/supporting electrolyte system for the following applications using the dc polarography technique.

(*i*)   Determination of Cd(II) in an approximately 0.001 mol dm^{-3} aqueous solution of $Cd(NO_3)_2$.

(*ii*)   Determination of nitrobenzene at the 10 ppm level in methanol.

(*iii*)   Determination of Pb(II) at a concentration of about 10^{-6} mol dm^{-3} in an aqueous solution.

SAQ 1.4a

SAQ 1.4b   What are the two roles of the supporting elec-
                  trolyte in dc polarography?

                  Calculate the relative magnitudes of the migra-
                  tion current and the diffusion current for the
                  zinc cation in an aqueous solution at 25 °C con-
                  taining 5×10^{-4} mol dm^{-3} Zn(NO$_3$)$_2$ and 0.1
                  mol dm^{-3} KCl.

                  Mobility values /m^2 V^{-1} s^{-1}:

                  NO$_3^-$ 7.4×10^{-8}; K$^+$ 7.6×10^{-8};

                  Cl$^-$ 7.9×10^{-8}; Zn^{2+} 5.5×10^{-8}.

SAQ 1.4b

SUMMARY AND OBJECTIVES

Summary

This section has been devoted to discussing the factors to be taken into account in choosing the solvent and supporting electrolyte. The dual role of the supporting electrolyte is established and in particular its role in ensuring diffusion control as the dominant mechanism for mass transport of analyte to the electrode illustrated. The marginal effect of temperature has been explained.

Your attention has been drawn to the fact that you now have available almost all of the information necessary to carry out a dc polarography experiment.

Objectives

You should now be able to:

- discuss the available solvents and the factors that determine the choice of solvent;

- explain the roles of the supporting electrolyte in dc polarography and give examples of supporting electrolytes used in protic and aprotic solvents;

- carry out calculations to illustrate the effect of excess supporting electrolyte on ion transport and to determine the electrical resistance of such solutions;

- select a suitable solvent/supporting electrolyte system for a particular analyte;

- state the effect of temperature on the results of a dc polarography analysis.

1.5.   THEORY AND FORM OF THE CURRENT/POTENTIAL RELATIONSHIP

This section introduces you to the factors that determine the overall rate of an electrode reaction in a system which is not stirred. This allows predictions of the shape of the resulting current/WE potential curves for a system which is under diffusion control. The work of Ilkovic in 1934 in deriving the current/analyte concentration relationship for the DME is covered and the Ilkovic equation is stated and partially derived. The Heyrovsky-Ilkovic equation (1935) is then derived; this provides an explanation of the shape of the current WE potential curve. This curve now becomes a polarogram and the half-wave potential is defined and related to the polarogram. Finally the question of the reversibility of the electrode reaction is discussed and tests for reversibility are given.

1.5.1.   General Factors Affecting the Current/WE Potential Relationship

In any electrolysis cell (Fig. 1.2a) the anode receives electrons from the solution (oxidation occurs) and the cathode receives electrons from the external voltage source (reduction occurs).

When an electrode is in equilibrium with a solution, with no external applied voltage, the electrode assumes a potential (E_e), the reversible electrode potential.

∏   For a reaction:

ox + ne → red,

what is the relationship between E_e and the concentration of oxidised (ox) and reduced (red) states in the solution?

This is asking you to recall the Nernst equation readily derivable from basic thermodynamic equations,

$$E_e = E^\ominus - (RT/nF) \ln c(\text{red})/c(\text{ox})$$

We have written this reversible potential as E_e to distinguish it from a general electrode potential, E. Values of $E^\ominus$, the standard electrode potential are given in Fig. 1.2c, and are always quoted on the NHE scale.

∏   For an aqueous solution of cadmium sulphate (0.1 mol dm^{-3}), $E^\ominus(Cd^{2+}, Cd) = -0.402$ V (NHE) at 25 °C. What is the value of $E_e(Cd^{2+}, Cd)$ on the SCE scale? 2.303 (RT/F) = 0.06 V at 25 °C.

Answer is -0.626 V (SCE). You should have arrived at this answer as follows.

E_e (NHE) $= E^\ominus - (0.06/2) \log (1/0.1)$

$= -0.402 - 0.03$

$= -0.432$ V

E_e (SCE) $= E_e$ (NHE) $- 0.244$ V

∴ $E_e = -0.676$ V (SCE)

The Faradaic current is a measure of the rate at which an electro-chemical reaction occurs and this rate is determined by two factors:

(a) the rate of the overall electron transfer process at the electrode surface, and

(b) the rate of movement of the electroactive species through the solution to the electrode – the rate of mass transport.

Considering each of these factors in turn.

(a) The electron transfer process.

The current at the working electrode (WE), I, is given by

$$I = I_a + I_c$$

where the relative magnitudes of I_a (anodic current) and I_c (cathodic current) reflect the extent to which oxidation and reduction are occuring at a particular electrode potential (E).

∴   When $E = E_e$, $I_a = I_c$ and $I = 0$, zero or null current conditions.

By convention the anodic current, I_a, is negative and the cathodic current, I_c, is positive.

When an external voltage is applied (E_{app}), the electrode potential changes from E_e to E and a finite current develops.

Thus when $E_{app} > 0$, $I_a \neq I_c$, $I \neq 0$, $I = f(E)$.

In Fig. 1.5a (i) we see they typical I/E curves for a system where the electron transfer process is very fast. We see that as E moves away from E_e the current is either pure anodic or pure cathodic. These fast electron transfer processes occur when the activation energy barrier to the reaction is low, a situation very common in the metal/metal cation reactions frequently encountered in dc polarography.

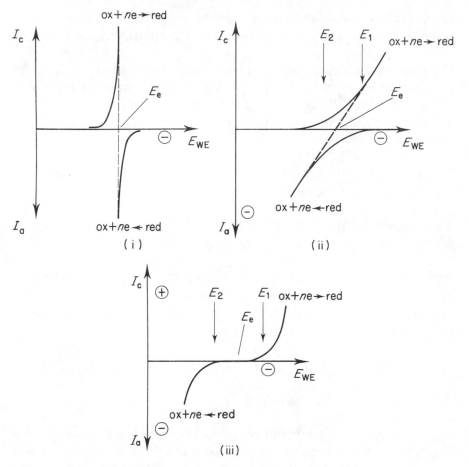

Fig. 1.5a.   *I/E curves for a pure electron transfer mechanism*
(- - - net current)

(*i*)   very fast electron transfer;
(*ii*)   slow electron transfer;
(*iii*) very slow electron transfer.

In Fig. 1.5a (*ii*) we see the typical I/E curves for a system where the electron transfer process is slow due to a high activation energy barrier for the reaction.

At $E = E_1, |I_c| > |I_a|$, thus I is positive.

At $E = E_2, |I_c| < |I_a|$, thus I is negative.

In general for a slow electron transfer process there is a wide range of potential where the current is of a mixed anodic/cathodic origin.

If the activation energy barrier is very high then a situation arises as depicted in Fig. 1.5a (*iii*). The I/E curves are completely separated and it is not until E exceeds E_2 in the anodic direction or E exceeds E_1 in the cathodic direction that any reaction occurs and hence any current is observed.

The potential difference $(E_1 - E_e)$ is the activation overpotential for the cathodic process and is negative.

The potential difference $(E_2 - E_e)$ is the activation overpotential for the anodic process and is positive.

Reactions which fall into the categories illustrated in Fig. 1.5a (*i*) and (*ii*) are said to be reversible in the electrochemical sense. Reactions of the type illustrated in Fig. 1.5a (*iii*) are said to be irreversible.

∏   Explain what you understand by activation overpotential.

The essential points are:

— if an electron transfer process is very fast then the potential at which the process occurs will be E_e, the reversible potential;

— if the activation energy barrier is high, additional potential is needed to achieve a finite rate and hence current;

— overpotential, η, is a measure of the additional potential required,

$$\eta = |E - E_e|$$

It has been assumed that you already have a knowledge of overpotential and its origins. If not you are advised to read the relevant sections in ACOL: *Principles of Electroanalytical Methods* or in Crow (1974).

(*b*) The mass transfer process.

In (*a*) we assumed a pure electron transfer controlled process at all potentials. As the electron transfer process becomes faster as a consequence of a more favourable electrode potential (eg more negative for a reduction), a situation will eventually arise where the electroactive material is unable to reach the electrode at a sufficiently fast rate. We then find that the current reaches a limiting value dependent upon the rate of mass transport.

There are three mass transport mechanisms capable of transferring electroactive material to and from the electrode:

(*i*)   migration under the potential gradient;

(*ii*) diffusion under the concentration gradient;

(*iii*) convection due to stirring and/or thermal agitation.

When a solution is quiescent (no stirring) we may assume that only factors (*i*) and (*ii*) are important. We have seen already (1.4) that steps are taken in dc polarography to ensure that factor (*i*) is negligible.

∏   How would you ensure that diffusion becomes rate-limiting and hence the current-limiting process is an electrochemical reaction?

      This is done by adding an excess of a supporting electrolyte to the solution. (Revise 1.4.2 if you do not understand the principle involved.)

If diffusion does become the rate-limiting process as in dc polarography then the relationship between current and overpotential changes. The characteristic feature of this relationship is the approach to a limiting current, I_{lim}, where

$$I_{lim} = nFDcA/\delta \qquad (1.5a)$$

In this equation $D/\text{m}^2\,\text{s}^{-1}$ is the diffusion coefficient of the analyte, $c/\text{mol m}^{-3}$ the analyte concentration, A/m^2 the electrode area, and δ/m the thickness of the electrical double layer. We shall take this up in greater detail (1.5.2) but it is sufficient here that the significance of δ alters for each model of the electrode/solution interface. Having considered the electron transfer and mass transport process separately, let us now consider them together.

(*c*) Electron transfer and mass transport.

The curves in Fig. 1.5a are idealised in the sense that mass transport was assumed to be infinitely fast at all potentials and could thus be ignored as a factor in limiting the process.

Fig. 1.5b shows the I/E curves for a cathodic process when the behaviour is determined by both electron transfer and mass transport. We see that the limiting current is the same for both fast and slow electron transfer processes but the slower the electron transfer process the more negative the potential required to achieve the limiting value.

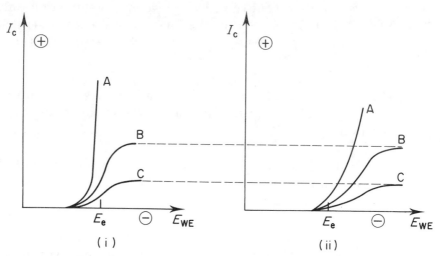

Fig. 1.5b.   *I/E curves for a cathodic electrode process*

(*i*)   very fast electron transfer,
(*ii*)   slow electron transfer.
(A ignoring mass transport, B fast mass transport and C slow mass transport).

Does this make sense to you? For a cathodic process as the electrode potential becomes more negative, even a naturally slow electron transfer process becomes faster and eventually the mass transport limit is reached.

How does all this relate to analysis? We now understand the factors affecting the shape of I/E curves when mass transport is diffusion controlled. The key to a successful analytical method will be if the current at some point on the I/E curve is reliably dependent upon the concentration of the analyte in the bulk solution. Ideally the relationship between current and concentration should be linear. Dc polarography is such a method.

SAQ 1.5a	State the main contributions to the Faradaic current in an electrolysis cell. How does diffusion control manifest itself in the current/working electrode potential relationship?

SAQ 1.5b Show by a sketch the effect of slow and fast electron transfer and slow and fast mass transfer upon the shape of I/E_{WE} curves.

1.5.2. The Ilkovic Equation

We have seen already that under the conditions of a typical dc polarography experiment the analyte approaches the spherical mercury drop under diffusion control. An equation is required which relates the current developed by this diffusion mechanism to the analyte concentration in the bulk solution. Such an equation has been avilable for diffusion to a planar electrode since 1902. It was derived by Cottrell and takes the form,

$$I_{\lim} = nFAcD^{\frac{1}{2}}/\pi^{\frac{1}{2}}t^{\frac{1}{2}} \qquad (1.5b)$$

Compare this with Eq. 1.5a. We see that in this situation δ has taken on the significance of $(\pi D t)^{\frac{1}{2}}$.

Ilkovic followed the same method as Cottrell but used as a model a growing spherical drop. This derivation is beyond the level of this unit so we will take a short cut and arrive at an equation resembling the Ilkovic equation by simply substituting the geometrical properties of a sphere into the Cottrell equation.

We have seen already (1.3.3) that for the DME,

$$A = (4\pi)^{\frac{1}{3}} (3)^{\frac{2}{3}} \rho^{-\frac{2}{3}} m^{\frac{2}{3}} t^{\frac{2}{3}}$$

Thus substituting into Eq. 1.5b,

$$I_{\lim} = \left(\frac{nFcD^{\frac{1}{2}}}{\pi^{\frac{1}{2}} t^{\frac{1}{2}}} \right) (4\pi)^{\frac{1}{3}} (3)^{\frac{2}{3}} \rho^{-\frac{2}{3}} m^{\frac{2}{3}} t^{\frac{2}{3}}$$

$$= 464 \, nD^{\frac{1}{2}} m^{\frac{2}{3}} t^{\frac{1}{6}} c$$

assuming $F = 96\,485$ C; $\rho(\text{Hg}) = 13\,534$ kg m^{-3}

The result includes an in-built error due to a neglect of the effect of the increase of drop size on the thickness of the diffusion layer. As the drop grows the diffusion layer stretches and thins, thus increasing the concentration gradient and the current. This factor was included in the Ilkovic derivation and it amounts to a correction of $(7/3)^{\frac{1}{2}}$.

$$\therefore \quad I_{\lim} = 708 \, nD^{\frac{1}{2}} m^{\frac{2}{3}} t^{\frac{1}{6}} c$$

In fact the Ilkovic equation derived in 1934 takes the form, for a cathodic process

$$I = 708 n D_{\text{ox}}^{\frac{1}{2}} m^{\frac{2}{3}} t^{\frac{1}{6}} (c^{\text{ox}} - c_0^{\text{ox}}) \tag{1.5c}$$

which gives the current at any potential, with c^{ox} being the bulk electrolyte concentration and c_0^{ox} the concentration at the electrode surface. When the potential is sufficiently negative (for a cathodic process),

$$c_0^{ox} \rightarrow 0, \text{ and } I \rightarrow I_{lim}.$$

This limiting current is called the diffusion current, I_d.

$$\therefore \qquad\qquad I_d = 708 \, nD_{ox}^{\frac{1}{2}} \, m^{\frac{2}{3}} \, t^{\frac{1}{6}} \, c^{ox} \qquad\qquad (1.5d)$$

The units are: $D/m^2 \, s^{-1}$; $c/mol \, m^{-3}$; $m/kg \, s^{-1}$; t/s (drop time) I_d/A.

Fig. 1.5c gives values of the diffusion coefficient of selected ions in the common supporting electrolyte, 0.1 mol dm^{-3} KCl at 25 °C. In general diffusion coefficients in aqueous solution lie in the range 7×10^{-10} to $20 \times 10^{-10} \, m^2 \, s^{-1}$ with the exception of the ions H_3O^+, and OH^-, which have much larger values.

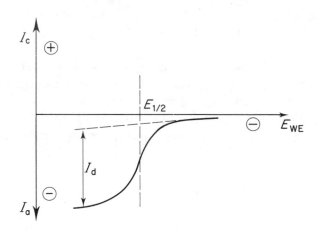

Ion	Zn^{2+}	Cd^{2+}	Pb^{2+}
$10^{10} \, D/m^2 \, s^{-1}$	6.73	7.15	8.67

Fig. 1.5c.   *Diffusion coefficients at 25 °C in 0.1 mol dm^{-3} aqueous KCl*

∏ Use the data in Fig. 1.5c together with the following data to calculate I_d for the Zn(II) ion in 0.1 mol dm^{-3} aqueous KCl at 25 °C, if the Zn(II) concentration is 10^{-3} mol dm^{-3}, $m = 0.002$ g s^{-1}, $t = 4$ s.

Answer is 7.35×10^{-6} A

Your only problem is likely to be one of using the correct units. If you use SI units consistently in Eq. 1.5d:

$$I_d = 708 \times 2 \times (6.73 \times 10^{-10})^{\frac{1}{2}} \times (2 \times 10^{-6})^{\frac{2}{3}} \times 4^{\frac{1}{6}} \times 1.$$

$$= 7.35 \times 10^{-6} \text{ A} = 7.35 \ \mu\text{A}$$

The assumptions that are implicit in the Ilkovic equation are:

— the flow rate of mercury is constant;

— the drops are spherical;

— there is no shielding of the drop by the capillary;

— the concentration of the analyte at the electrode surface is zero when the limiting current is achieved;

— the solution is not stirred;

— the linear diffusion theory developed by Cottrell is valid.

None of these assumptions is absolutely correct but more advanced theories produce equations very close to that of Ilkovic and with the same concentration dependence.

∏ You have learned that both m and t are dependent upon the height of the mercury column (h) in the DME. Using this knowledge (1.3.2) derive the dependence of I_d upon h.

Answer is $I_d \propto h^{\frac{1}{2}}$

Previously you learned, $m \propto h$ and $t \propto h^{-1}$.

$$\therefore \quad I_d \propto m^{\frac{2}{3}} \, t^{\frac{1}{6}} \propto h^{\frac{2}{3}} \, h^{\frac{1}{6}} \propto h^{\frac{1}{2}}$$

Consider the time dependence of the diffusion current,

$$I_d \propto t^{\frac{1}{6}}$$

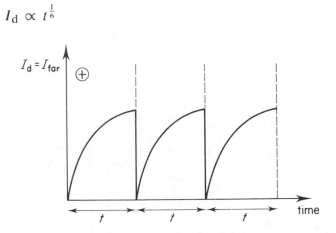

Fig. 1.5d.   *Diffusion current as a function of t
(drop lifetime)*

Fig. 1.5d illustrates this relationship. A good quality pen recorder (1 s response) is capable of following these oscillations. Earlier reports of dc polarography results used instruments that produced a highly damped current output. This effectively records the current averaged over the drop lifetime whereas the Ilkovic equation above gives the diffusion current at the instant that the drop falls. The mean or average diffusion current $(\bar{I}_d)$ is given by

$$\bar{I}_d = 607 \, n \, D_{\text{ox}}^{\frac{1}{2}} \, m^{\frac{2}{3}} \, t^{\frac{1}{6}} \, c^{\text{ox}} \tag{1.5e}$$

The capacitive current (I_{cp}) is given by Eq. 1.3i:

$$I_{\text{cp}} = k' \, C_{\text{DL}} \, E(\text{PZC}) \, t^{-\frac{1}{3}} \tag{1.3i}$$

and the time dependence of this current is shown in Fig. 1.3d. Under normal experimental conditions, the Faradaic current, measured by

I_d, far exceeds the capacitive current and the total current behaves as in Fig. 1.5d. As the concentration of analyte falls to the 10^{-4} -10^{-5} mol dm^{-3} level so the two types of current (I_d and I_{cp}) approach each other in value and a time dependence as shown in Fig. 1.5e is obtained.

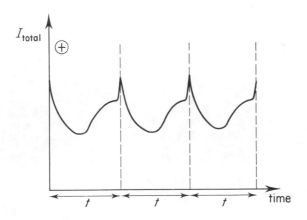

Fig. 1.5e. *Variation of total Faradaic + capacitive current with t (drop lifetime) when $I_{far} \approx I_{cp}$*

1.5.3. The Heyrovsky–Ilkovic Equation

This equation may take a variety of forms depending upon the behaviour of the oxidised and reduced species at the electrode. The version presented here is applicable to a situation where the oxidised species is the only species in the solution and this species is being reduced at a cathode where the reduced species either goes into solution or forms an amalgam with mercury. This equation is then directly relevant to most of the applications of dc polarography. Consider the Ilkovic equation applied to the reaction:

$$ox + ne \rightarrow red$$

$$I = 708 \ nFD_{ox}^{\frac{1}{2}} \ m^{\frac{2}{3}} t^{\frac{1}{6}}(c^{ox} - c_o^{ox}) = k \ (c^{ox} - c_o^{ox}) \qquad (1.5f)$$

Similarly for the reaction:

$$\text{red} \rightarrow \text{ox} + ne$$

$$I = 708\, nFD_{\text{red}}^{\frac{1}{2}}\, m^{\frac{2}{3}}\, t^{\frac{1}{6}}\, (c_0^{\text{red}} - c^{\text{red}}) = k'c_0^{\text{red}} \tag{1.5g}$$

Note the reversal of the concentration expression and the fact that c^{red} is here zero (only the oxidised species is present in the bulk solution).

Apply the Nernst equation to the electrode surface,

$$E = E^{\ominus} - \frac{RT}{nF} \ln \frac{c_0^{\text{red}}}{c_0^{\text{ox}}} \tag{1.5h}$$

Substituting from 1.5f and 1.5g for $c_0\text{ox}$ and $c_0\text{red}$ into 1.5h we obtain:

$$E = E^{\ominus} - \frac{RT}{nF} \ln \frac{(I/k')\, k}{(kc^{\text{ox}} - I)}$$

Now $I_{\text{d}} = kc^{\text{ox}}$,

$$\therefore \quad E = E^{\ominus} - \frac{RT}{nF} \ln \frac{kI}{k'(I_{\text{d}} - I)}$$

Looking at the significance of k and k' in 1.5f and 1.5g we have,

$$E = E^{\ominus} - \frac{RT}{nF} \ln \left\{ \frac{D_{\text{ox}}^{\frac{1}{2}}}{D_{\text{red}}^{\frac{1}{2}}} \times \frac{I}{(I_{\text{d}} - I)} \right\}$$

$$\therefore \quad E = E^{\ominus} - \frac{RT}{nF} \ln \left\{ \frac{D_{\text{ox}}^{\frac{1}{2}}}{D_{\text{red}}^{\frac{1}{2}}} \right\} - \frac{RT}{nF} \ln \left\{ \frac{I}{I_{\text{d}} - I} \right\} \tag{1.5i}$$

You should note that if we are considering in detail this derivation we should have used activities and not concentrations in the Nernst equation (1.5h)

ie $\quad E = E^{\ominus} - \dfrac{RT}{nF} \ln \dfrac{a_0^{\text{red}}}{a_0^{\text{ox}}}$

$$= E^{\ominus} - \frac{RT}{nF} \ln \frac{\gamma_0^{\text{red}}}{\gamma_0^{\text{ox}}} - \frac{RT}{nF} \ln \frac{c_0^{\text{red}}}{c_0^{\text{ox}}}$$

where γ_0^{red} and γ_0^{ox} are respectively the activity coefficients of the reduced and oxidised species at the electrode surface. This change leads to a modification of Eq. 1.5i to give,

$$E = E^{\ominus} - \frac{RT}{nF} \ln \left\{ \frac{\gamma_0^{\text{red}} D_{\text{ox}}^{\frac{1}{2}}}{\gamma_0^{\text{ox}} D_{\text{red}}^{\frac{1}{2}}} \right\} - \frac{RT}{nF} \ln \left\{ \frac{I}{I_{\text{d}} - I} \right\} \qquad (1.5j)$$

This is the Heyrovsky–Ilkovic equation first derived in 1935.

This gives significance to an important analytical quantity the half-wave potential, $E_{\frac{1}{2}}$.

We may define

$$E_{\frac{1}{2}} = E^{\ominus} - \frac{RT}{nF} \ln \left\{ \frac{\gamma_0^{\text{red}} D_{\text{ox}}^{\frac{1}{2}}}{\gamma_0^{\text{ox}} D_{\text{red}}^{\frac{1}{2}}} \right\} \qquad (1.5k)$$

Note that the second term is usually close to zero, hence $E_{\frac{1}{2}} \approx E^{\ominus}$.

Thus from Eq. 1.5j:

$$E = E_{\frac{1}{2}} - \frac{RT}{nF} \ln \left\{ \frac{I}{I_{\text{d}} - I} \right\} \qquad (1.5l)$$

Putting $I = I_{\text{d}}/2$, we have $E = E_{\frac{1}{2}}$ which provides an alternative definition of the half-wave potential. It is the potential at which the Faradaic current (I) is one half of the diffusion current I_{d}. The two definitions agree provided the reaction is reversible.

Fig. 1.5f shows a plot of Eq. 1.5l in the form of $I = f(E)$. This is called a polarogram and the shape is a polarographic wave. The residual current is mainly due to the capacitive current.

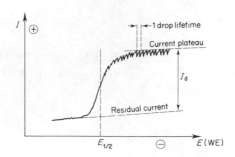

Fig. 1.5f.   *Typical polarogram, a polarographic wave*

1.5.4.   Reversibility

We have discussed (1.5.1) the two factors, electron transfer and mass transport, which determine the rate of an electrode reaction. Figs. 1.5a and 1.5b illustrate the current/WE potential curves obtained for reactions which are said to be either reversible or irreversible. The basic factor which determines whether or not an electrode reaction is reversible is the rate of the electron transfer process. This is measured in terms of a rate constant (k^o) and in general if:

$$k^o > 2 \times 10^{-2} \text{ cm s}^{-1} \text{ the reaction is reversible;}$$

while if

$$k^o < 2 \times 10^{-5} \text{ cm s}^{-1} \text{ the reaction is irreversible.}$$

The rate constant for the diffusion process (k_d) has a value which remains fairly constant for most aqueous solution processes at about 1×10^{-3} cm s^{-1}. We see that for a reversible process, $k^o \gg k_d$.

Fig. 1.5g shows polarograms for reversible and irreversible processes. It is important to note that the diffusion current limit (I_d) is still attained but not at the same potential. This is an important point in analytical applications. Note however that the significance of the half-wave potential ($E_{\frac{1}{2}}$) has changed considerably. Theories of the irreversible process show that $E_{\frac{1}{2}}$ is now a function of k^o, D, γ and t but this is beyond the scope of this Unit.

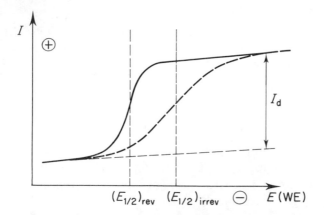

Fig. 1.5g. *Polarographic waves for a reversible (──) and an irreversible (- - - -) process. (Curves smoothed to remove oscillations)*

Tests for Reversibility

There are two tests which are in common use. Both require appropriate data to be read off an experimental polarogram.

(i) This involves the direct use of Eq. 1.5l. A plot is made of log $I/(I_d - I)$ against E Fig. 1.5h.

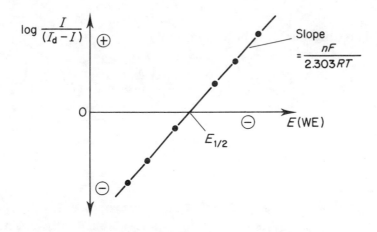

Fig. 1.5h. *Test for reversibility*

The slope of the line is $-(nF/2.30RT)$, the 2.303 arising from the conversion from natural logarithms to logarithms to base 10. The intercept when $\log I/(I_d - I) = 0$, gives the value of $E_{\frac{1}{2}}$.

Fig. 1.5i gives values of $(2.303RT/F)$ at various temperatures. For a reaction to be reversible you should obtain a straight line with a slope close to the theoretical value. As the reaction moves towards irreversible conditions, a straight line is still obtained but the value of the slope is not the theoretical one.

$T/°C$	17	19	21	23	25	27	29
$2.303RT/F/mV$	57.6	58.0	58.4	58.7	59.1	59.5	59.9

Fig. 1.5i.   *Values of 2.303 RT/F at different temperatures*

(ii) The second method, attributed to Meites, is much simpler. The values of the potentials ($E_{\frac{1}{4}}$ and $E_{\frac{3}{4}}$) are read from the polarogram where I is $1/4$ and $3/4$ of I_d.

The difference of these potentials

$$E_{\frac{3}{4}} - E_{\frac{1}{4}} = -\frac{0.0564}{n} \text{ V at 25 °C} \tag{1.5m}$$

∏   Try to prove the Meites relationship by using the Heyrovsky–Ilkovic equation at 25 °C.

The method is as follows:

$$E = E_{\frac{1}{2}} - \frac{0.0591}{n} \log \frac{I}{I_d - I}$$

$$\therefore \qquad E_{\frac{1}{4}} = E_{\frac{1}{2}} - \frac{0.0591}{n} \log \left\{ \frac{I_\mathrm{d}/4}{I_\mathrm{d} - I\mathrm{d}/4} \right\}$$

$$= E_{\frac{1}{2}} - \frac{0.0591}{n} \log 1/3$$

Similarly, $\quad E_{\frac{3}{4}} = E_{\frac{1}{2}} - \frac{0.0591}{n} \log 3$

$$\therefore \qquad E_{\frac{3}{4}} - E_{\frac{1}{4}} = - \frac{0.0591}{n} (\log 3 - \log 1/3)$$

$$= - \frac{0.0564}{n} \text{ V}$$

For an irreversible system $(E_{\frac{3}{4}} - E_{\frac{1}{4}})$ is still negative but numerically greater than the value given in Eq. 1.5m.

∏   The following measurements were made on a polarographic wave at 25 °C for the reaction:

ox + 2e → red,

for which $I_\mathrm{d} = 3.24 \ \mu\text{A}$.

E/V(SCE)	$I/\mu\text{A}$
−0.395	0.48
−0.406	0.97
−0.415	1.46
−0.422	1.94
−0.431	2.43
−0.445	2.92

Show that this is a reversible process and calculate $E_{\frac{1}{2}}$.

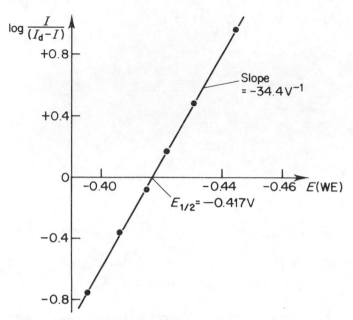

Fig. 1.5j.   *Application of test for reversibility*

The answer is seen in Fig. 1.5j where log $I/(I_d - I)$ is plotted against E. The process is reversible because the measured slope (-34.4 V^{-1}) is close to the theoretical value of $2F/2.303RT$ (ie -33.8 V^{-1}). $E_{\frac{1}{2}} = -0.417$ V.

1.5.5.   Anodic and Mixed Anodic/Cathodic Waves

We said at the beginning of 1.5.3 that the Heyrovsky–Ilkovic equation would be derived for a pure cathodic wave and this version of the equation took the form:

$$E = E_{\frac{1}{2}} - \frac{RT}{nF} \ln\left\{\frac{I}{(I_d)_c - I}\right\} \qquad (1.5n)$$

where $(I_d)_c$ is the diffusion current for a cathodic wave. A similar equation may be derived for a pure anodic wave, ie the situation where the only species present in solution is the reduced species and the reaction, red → ox + *n*e,

occurs. Both oxidised and reduced species are assumed to be soluble either in the solution or in the mercury drop. This equation takes the form:

$$E = E_{\frac{1}{2}} - \frac{RT}{nF} \ln\left\{\frac{(I_d)_a - I}{I}\right\} \qquad (1.5o)$$

where $(I_d)_a$ is the diffusion current for an anodic wave. These are both specific cases of the more general equation which is applicable to a situation where both oxidised and reduced species, eg Fe(III)/Fe(II), are present in the solution and both are soluble species. This general equation is:

$$E = E_{\frac{1}{2}} - \frac{RT}{nF} \ln\left\{\frac{(I_d)_a - I}{I - (I_d)_c}\right\} \qquad (1.5p)$$

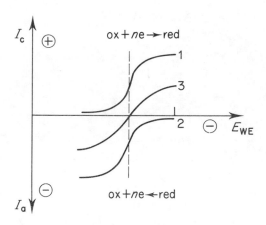

Fig. 1.5k. *Anodic/cathodic waves for a reversible system*

  1. cathodic wave (only ox present);
  2. anodic wave (only red present);
  3. both ox and red present (equal concentrations)

These relationships are illustrated in Fig. 1.5k for a system where the concentration of anodic and cathodic species are equal. Fig. 1.5l shows the expected mixed wave when the two species have different concentrations.

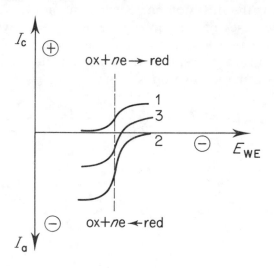

Fig. 1.5l.   *Anodic/cathodic waves for a reversible system, unequal concentration*

1. cathodic wave (only ox present);
2. anodic wave (only red present);
3. both ox and red present ($c^{red} > c^{ox}$)

The question of reversibility has been discussed (1.5.4) The Heyrovsky–Ilkovic equation cannot be applied to an irreversible process but the trends shown in Figs. 1.5k and 1.5l are repeated as seen in Fig. 1.5m.

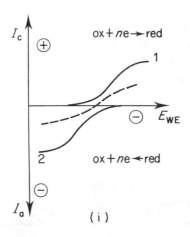

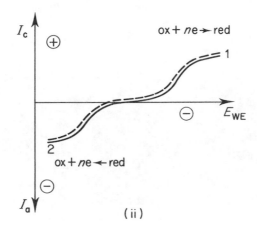

Fig. 1.5m. *Anodic/cathodic waves for (i) an irreversible process and (ii) a very irreversible process*

      1. cathodic wave,

      2. anodic wave,

      - - - - mixed wave.

SAQ 1.5c

State the general form of the Ilkovic equation for a reduction process and explain the terms used. From this equation derive the form of the Heyrovsky-Ilkovic equation that applies to the reaction:

$$red \rightarrow ox + ne$$

where both reduced and oxidised species are soluble, and only the reduced species is present in the solution. Draw the wave obtained in this situation.

SAQ 1.5d   What is the main factor that determines the limit of detection in dc polarography? Sketch the appearance of the polarograms of a typical supporting electrolyte and a typical analyte solution when the analyte concentration is $>10^{-3}$ mol dm^{-3} (including the shape of the oscillations in current due to drop size).

SAQ 1.5e

Use a graphical method and the Meites method to determine whether or not the following data relates to a reversible reaction. What is the significance of $E_{\frac{1}{2}}$ for a reversible reaction; is this significance valid for this reaction?

E_{WE} (SCE)/V: −0.419 −0.451 −0.491 −0.519 −0.561

$I/\mu A$: 0.31 0.62 1.24 1.86 2.48

The reaction is

$$ox + 2e \rightarrow red$$

$I_d = 3.10 \ \mu A$; temperature 25 °C.

SAQ 1.5f   If you had a mixture of Sn(IV)/Sn(II) ions in solution and the concentration ratio of Sn(IV): Sn(II) was 3 : 1 sketch the appearance of the polarogram expected assuming that the reaction $Sn^{4+} + 2e \rightleftarrows Sn^{2+}$ is reversible.

SUMMARY AND OBJECTIVES

Summary

The factors; rate of electron transfer and rate of mass transport are discussed, and the resulting I/E(WE) curves given for reversible and irreversible systems. The effect of diffusion control in imposing a limiting current is explained.

The application of the diffusion theory to the DME/solution interface leads to the Ilkovic equation relating the diffusion current (I_d) to the bulk analyte concentration. The Heyrovsky-Ilkovic equation leads directly to the shape of a typical polarogram (I/E(WE) curve) for a reversible process, and hence to the half-wave potential ($E_{\frac{1}{2}}$). Reversible and irreversible electrode reactions are considered and tests for reversibility given. The possibility of obtaining anodic waves and mixed waves in addition to cathodic waves is discussed.

Objectives

You should now be able to:

- explain the origin of a limiting current in a diffusion controlled electrode reaction;

- illustrate by sketches the effect of the rates of electron transfer and mass transport on the shapes of I/E curves;

- state the Ilkovic equation; explain its origins and the assumptions upon which it is based and calculate the diffusion current using the equation;

- sketch the appearance of the diffusion current/time curve during the drop lifetime;

- derive the Heyrovsky–Ilkovic equation for a cathodic process and use it to explain the half-wave potential and the shape of the I/E(WE) curve;

- sketch the appearance of a typical polarogram and label the sketch;

- explain the meaning of reversible and irreversible in the context of electrode reactions and apply tests for reversibility;

- explain the origin of anodic and mixed waves and sketch the appearance of such waves for reversible and irreversible systems.

1.6. QUALITATIVE AND QUANTITATIVE ANALYSIS OF METALLIC CATIONS

We have seen in the previous sections details of the required circuitry, the design of the cell, the solution conditions and the theory relating observed current to analyte concentration. In this section we bring all of this together and apply it to the analytical problem. The treatment is deliberately restricted to applications to metallic cation analysis and only at the end is brief reference made to other applications.

1.6.1. Summary of Typical Experimental Conditions

It is appropriate here to bring together the knowledge of the experimental details of dc polarography that you have gained in the earlier sections. Briefly, in order to carry out the analysis for a metallic cation in solution, eg Cu(II), you would need to:

— set up the circuit (Fig. 1.2d), the components of which apart from the cell, would be incorporated in a modern commercial polarograph;

— assemble a three-electrode cell (Fig. 1.2g) using a DME (WE), Pt (SE) and SCE (RE);

— ensure that designed into the cell there were facilities for deoxygenating the solution and thermostatting the solution if necessary;

— introduce the analyte in the concentration range 10^{-2}–10^{-5} mol dm^{-3} in a suitable solvent;

— incorporate into the solution a supporting electrolyte of concentration at least 10 times greater than the analyte concentration;

— ensure that a very small amount of maximum suppressor is added to the solution, eg Triton X-100;

— deoxygenate the solution if a reduction process is to be studied;

— check that the DME is providing drops of mercury at a suitable rate or actuate the mechanical device for tapping the electrode;

— using a quiescent solution scan the potential over the required range of potential at a rate of 2–10 mV s^{-1} and in the appropriate direction, eg 0 to -2.0 V(SCE) for reductions, -2.0 V to 0 V (SCE) for oxidations;

— record the resulting polarogram on a x–y recorder with a fast response time. A result similar to Fig. 1.5f will be obtained if a metallic cation is the analyte.

1.6.2. Interpretation of a Typical Result – Qualitative Analysis

We have seen that a typical dc polarogram for a metallic cation takes the form shown in Fig. 1.5f. We shall be looking at such polarograms on several occasions from now on and it is convenient to omit the oscillations which are a feature of such polarograms, especially on the current plateau. You should not forget that these oscillations exist and their origins. Fig. 1.6a repeats the presentation of a dc polarogram and illustrates the two ways of obtaining the residual current.

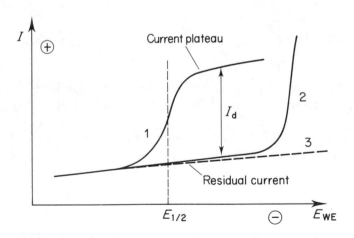

Fig. 1.6a.   *Typical polarogram of a metallic cation*

1. polarogram of analyte solution;
2. polarogram of solution containing
   only the supporting electrolyte;
3. extrapolated linear part of 1

It may be obtained either by recording a polarogram for the supporting electrolyte in the absence of the analyte or by extrapolating the
initial linear portion of the analyte polarogram. It is best, during a
series of experiments, to at least once verify that the two approaches
give the same result. The most likely cause of a difference is an electroactive impurity in the analyte. We have dealt at some length with
the capacitive current (I_{cp}) and you are reminded that this is expected to be the main contribution to residual current. Impurities
in the solvent, supporting electrolyte and analyte form the remainder and are usually negligible. Inadequate and/or variable levels of
deoxygenation are an important source of error.

(*a*) Measurement of half-wave potential and its significance.

Fig. 1.6b shows a dc polarogram and a simple geometric construction which provides a measure of objectivity in the assignment of
the value of $E_{\frac{1}{2}}$. This method works even if the residual current and
the current plateau lines are not parallel.

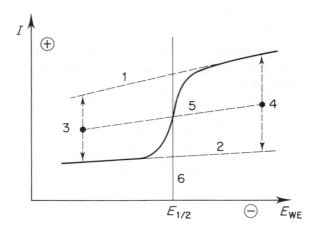

Fig 1.6b. *Determination of half-wave potential*

Extrapolated lines (1) and (2) are constructed followed by (3) and
(4) drawn vertically. The mid-points of (3) and (4) are now joined
by (5) and where (5) intersects the wave, a vertical line (6) is drawn
to locate $E_{\frac{1}{2}}$. This is a distinct improvement on trying to estimate
by eye the point of inflection on the wave.

∏   What other method do you know that gives an unambiguous
       value of $E_{\frac{1}{2}}$?

       In Section 1.5 we used a graphical test for reversibility based
       on the Heyrovsky–Ilkovic equation. For a reversible reaction
       this gives a value of $E_{\frac{1}{2}}$ from an intercept on the plot.

Having obtained a value of $E_{\frac{1}{2}}$ from the polarogram, what is the significance of this value? If we inspect the Heyrovsky-Ilkovic equation
we see (Eq. 1.5k) that:

$$E_{\frac{1}{2}} = E^{\ominus} - \frac{RT}{nF} \ln\left\{ \frac{\gamma_0^{red} D_{ox}^{\frac{1}{2}}}{\gamma_0^{ox} D_{red}^{\frac{1}{2}}} \right\}$$

All of the factors in this equation are properties of the oxidised or
reduced form of the analyte and we can say that $E_{\frac{1}{2}}$ is characteristic
of the reaction,

         ox + ne ⇌ red,

under the conditions of the experiment. The half-wave potential is
the quantity obtained from dc polarography that may be used for
qualitative analysis. The values are diagnostic for the cation which
is present. Fig. 1.6c gives values of $E_{\frac{1}{2}}$ for selected metallic cations
all in the same supporting electrolyte. Fig. 1.6d shows the effect of
a change in supporting electrolyte on the value of $E_{\frac{1}{2}}$.

Analyte:	O_2	Pb(II)	Tl(I)	Cd(II)	O_2	Zn(II)	H_3O^+
$E_{\frac{1}{2}}(SCE)/V$:	-0.05	-0.40	-0.46	-0.58	-0.90	-1.00	-1.58
$E^{\ominus}(SCE)/V$:	$-$	-0.36	-0.57	-0.60	$-$	-1.00	$-$

Fig. 1.6c. *Values of $E_{\frac{1}{2}}$ for metallic cations in 0.1 mol dm^{-3} aqueous KCl at 25 °C*

	Supporting electrolyte		
	1 mol dm^{-3} KCl	1 mol dm^{-3} KCN	0.5 mol dm^{-3} Na_2 tartrate, pH 9
Pb(II)	-0.44	-0.72	-0.58
Cd II	-0.64	-1.18	-0.64
Zn(II)	-1.00	-1.01	-1.15

Fig. 1.6d. *Effect of supporting electrolyte on the value of $E_{\frac{1}{2}}$ at 25 °C*

For a more comprehensive compilation of $E_{\frac{1}{2}}$ data you are referred to Meites (1965). We can use such tables of $E_{\frac{1}{2}}$ data in much the same way as characteristic absorption frequency tables are used in infrared spectroscopy; they indicate the likely content of the solution. Fig. 1.6e shows the polarogram expected for a solution containing lead, cadmium and zinc in aqueous 0.1 mol dm^{-3} KCl.

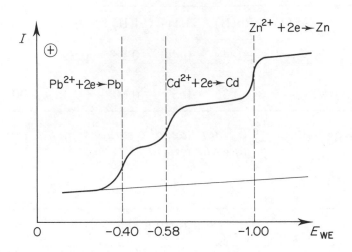

Fig. 1.6e.   *dc polarogram for a solution containing Pb(II), Cd(II) and Zn(II) in 0.1 mol dm^{-3} aqueous KCl*

(*b*) Multi-analyte analysis.

This leads us to consider the feasibility of multi-analyte analysis. Clearly if two waves are too close together there will be interference which will make accurate interpretation difficult. It is generally accepted that if the difference in half-wave potentials of successive waves ($\Delta E_{\frac{1}{2}}$) is >300/n mV, then the interference is negligible.

∏   Why does the value of n have a bearing on this criterion for lack of interference between successive waves?

       This is because n affects the shape of the polarographic wave. The Ilkovic equation tells us that there is a direct effect on the wave height, ie on I_d. The Heyrovsky-Ilkovic equation tells us that the value of n will affect the slope of the rising part of the wave. The greater the value of n the greater the slope, (Fig. 1.6f).

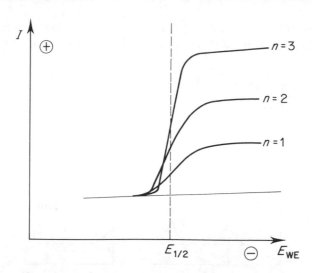

Fig. 1.6f. *Effect of n on the shape of a reversible polarographic
wave for the same concentration of analyte*

Those of you who wish can differentiate the Heyrovsky–
Ilkovic equation (1.5j) to obtain dE/dI and then invert this
to give dI/dE. Then put $I = I_d/2$, when $E = E_{\frac{1}{2}}$ and show
that:

$$dI/dE = -\left(\frac{nF}{RT}\right)I_d \qquad (1.6a)$$

Thus when $n = 3$ the slope is 9 times as great and the wave
height 3 times as large as when $n = 1$, for the same concentration of analyte.

There was available in the 1950's electronic circuitry that was capable of differentiating the observed current and presenting dI/dE as
a function of E. Fig. 1.6g illustrates this facility. $E_{\frac{1}{2}}$ emerges as the
peak maximum and the peak height at the maximum is proportional
to I_d. This is easily seen in Eq. 1.6a. A useful secondary advantage of
these derivative polarograms is that closely spaced waves are more
easily distinguished as peaks, ie the resolution is improved. This

simple extension of the dc method must not be confused with differential methods that will be described in later parts of this Unit.

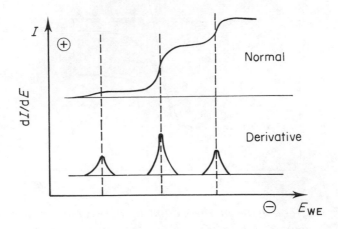

Fig. 1.6g.   *Derivative polarogram*

1.6.3.   Interpretation of a Typical Result, Quantitative Analysis

The Ilkovic equation for metallic cation analysis takes the form

$$I_{\mathrm{d}} \;=\; 708nFD_{\mathrm{ox}}^{\frac{1}{2}} \; m^{\frac{2}{3}} \; t^{\frac{1}{6}} \; c^{\mathrm{ox}} \tag{1.6b}$$

Quantitative analysis using dc polarography amounts to making use of the linear relationship between the diffusion current and the bulk analyte concentration. It may seem attractive to use the Ilkovic equation directly. However the diffusion coefficient (D_{ox}) is usually not known accurately for the particular solution conditions and more important, the constant 708 emerges from the less than perfect model of the diffusion layer used in the derivation. The errors introduced are usually < 10% but this is not satisfactory for serious analytical work.

We turn then to the well-established methods used throughout analytical chemistry for processing data to obtain a quantitative result. Two of these methods are used routinely in dc polarography.

(*a*) Direct use of a calibration or working curve.

Standard solutions having known concentrations of analyte are prepared and their dc polarograms are separately run to obtain values
of the diffusion current. A calibration or working curve is produced
($I_d = f(c^{ox})$) which is usually linear. The polarogram of the unknown is now run under identical conditions and from the value of
I_d, and using the calibration curve, c^{ox} (unknown) is determined
(Fig. 1.6h).

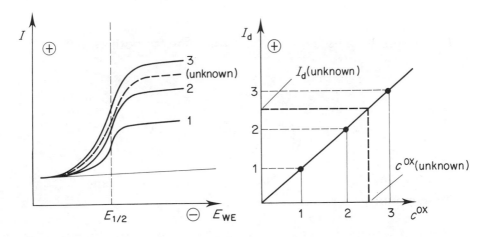

Fig. 1.6h. *Calibration curve in dc polarography*

It may not be immediately obvious from Section 1.2 but the assem-
bling of the cell and subsequent emptying, cleaning and re-filling
can be very time consuming. Hence dc polarography would benefit
from a method that obviated the need to empty and re-fill the cell.
Such a technique is known as the method of standard additions.
There are variants on this method but we will describe the one that
is most useful for our purpose. An alternative version using a graphical method is described in 5.6.2.

(*b*) The method of standard additions.

In this method the dc polarogram of the unknown is run and careful note is made of the volume of solution used (*v*). Let the mea-

sured diffusion current be $I_d(1)$. A standard solution of the analyte
is made of concentration S and a volume (V) of this standard is
added to the original solution. The mixture is stirred during the
deoxygenation stage and the new (increased) diffusion current is
measured – let this be $I_d(2)$. If the unknown analyte concentration
was x, then from the Ilkovic equation,

$$I_d(1) \propto x$$

and $$I_d(2) \propto \left(\frac{v}{V+v}\right)x + \left(\frac{V}{V+v}\right)S$$

The proportionality constants are the same for both equations.

Thus

$$\frac{I_d(2)}{I_d(1)} = \left(\frac{v}{V+v}\right) + \left(\frac{V}{V+v}\right)\frac{S}{x} \qquad (1.6c)$$

The experiment is usually designed so that $v \gg V$, and $S \gg x$. In
these circumstances,

$$\frac{I_d(2)}{I_d(1)} \approx 1 + \left(\frac{V}{v}\right)\frac{S}{x} \qquad (1.6d)$$

Π   A polarographic cell contains 10 cm^3 of a solution of analyte
     X in a supporting electrolyte and gives a diffusion current of
     $3.6 \ \mu\text{A}$. 0.10 cm^3 of a standard solution containing 5×10^{-2}
     mol dm^{-3} X is now added and the new polarogram yields
     an I_d value of $7.2 \ \mu\text{A}$. What is the concentration of X in the
     original solution?

     The answer is 5×10^{-4} mol dm^{-3}

     Using Eq. 1.6d we have:

$$\frac{7.2}{3.6} \approx 1 + \frac{(0.10)}{10}\frac{5 \times 10^{-2}}{x}$$

$$2 \approx 1 + \left(\frac{5 \times 10^{-4}}{x} \right)$$

$$x \approx 5 \times 10^{-4} \text{ mol dm}^{-3}$$

Using Eq. 1.6c the answer would have been $4.9 \times 10^{-4} \text{ dm}^{-3}$, a 2% difference. Show that if 1.0 cm^3 were added instead of 0.10 cm^3 the diffusion current would be $36.6 \ \mu A$ and that the difference between the two values for x calculated using Eq. 1.6c and 1.6d is about 11%.

Further additions may be made to the cell and each time the enhanced value of I_d should fit Eq. 1.6c but the validity of 1.6d would become progressively worse.

∏   What advantages does the use of the method of standard additions have over the direct use of a calibration curve, other than obviating the need to empty, clean and re-fill the cell?

This is testing your background knowledge of analytical chemistry. If there are present in the analyte unknown impurities that could interfere with the analytical result then the method of standard additions is to be preferred. The calibration curve is constructed using results from standards containing only the analyte. If interferences are present in the unknown an erroneous result will be obtained. In the method of standard additions the unknown is present for all the measurements, hence so too are the impurities. The interference, if present, is common to all the measurements.

This is a convenient moment to make another point. In the method described above we have made measurements on the same unknown at least twice and you have been told that repeated measurements may be made after successive additions of standard. Quite separately you should realise that one of the great advantages of dc polarography is that you can make repeated measurements on the same solution and obtain the same answer each time.

∏   Why is it that the measured concentration of analyte does not decrease?

Consider a typical dc polarography experiment. At a scan rate of 2 mV s^{-1}, a wave is completed in about 100 s. During this time the current varies from <1 μA to say 50 μA and it would be fair to say an average of 10 μA is obtained. For a 10^{-4} mol dm^{-3} aqueous solution of $Pb(NO_3)_2$ in 10 cm^3 of solution, what is the change in concentration as the result of the experiment?

Answer is approximately 0.5% difference.

10 μA for 100 s = $10 \times 10^{-6} \times 10^2 = 10^{-3}$C

Number of Faraday passed = $10^{-3}/96485 = 1.04 \times 10^{-8}$

1 Faraday deposits 0.5 mol Pb

∴ 5.2×10^{-9} mol Pb removed from solution

Solution had $10^{-4}/10^2$ mol Pb in 10cm^3 = 10^{-6} mol

∴   Change is 5.2×10^{-9} mol in 10^{-6} mol, ie $\approx$ 0.5%

Very little electrolysis occurs during any one experiment. Dc polarography is classified as an example of a microelectrolysis method for this reason and we see that it is also virtually non-destructive.

We have discussed quantitative analysis in terms of reversible reactions and you are reminded that irreversible reactions do not yield a value of $E_{\frac{1}{2}}$ with the same significance as that for reversible reactions. What about the value of I_d – is this of any value? The answer is yes, usually. The value of I_d obtained is still given by the Ilkovic equation and the linear relationship between I_d and c^{ox} is often retained. For any reaction you must measure I_d as a function of the concentration of analyte and investigate the relationship. Usually this is linear and analysis is carried out in the same manner as for a reversible reaction.

1.6.4. Analytes other than Metallic Cations

A brief mention only is made here of the wide range of applications of dc polarography. It is possible to study oxidation reactions as well as reduction reactions, provided they occur within the voltage limits set by the DME/solvent/supporting electrolyte system. Examples of metallic cation oxidations are Cu(I); Ce(III). Inorganic species, other than metallic cations may also be determined, eg NO_3^-, NO_2^-, S^{2-}. Finally there is the extensive and rapidly growing field of application to organic materials particularly in the biochemical, medicinal and pharmaceutical areas. Provided it is possible to oxidise or reduce an organic species chemically it is almost certain that the same reaction can be brought about electrochemically. Often the electrochemical method offers greater selectivity, ie control over the nature of the reaction product. Examples of organic analysis by polarography are: benzodiazepines (eg Valium); antibiotics (eg streptomycin); vitamins (eg riboflavin); steroids (eg progesterone). This is an impressive list and shows the versatility of the technique. Bear in mind that to reach the low detection limits required in realistic determinations of these compounds, the more advanced forms of polarography would be required.

Before leaving the subject you should be reminded again that dc polarography yields more than just an analytical result. It gives information about the rates of the electrode processes and is a very valuable tool for mechanistic studies, especially when the process becomes irreversible. In this area (non-analytical) the dc method remains a better method than most of the more advanced techniques.

| SAQ 1.6a | Describe a typical dc polarography experiment designed to obtain the polarogram of a mixture of Pb(II) and Zn(II) ions in their aqueous solutions as nitrates at 25 °C and both at concentrations of about 10^{-4} mol dm^{-3}. Sketch the expected polarogram and show how you would determine the value of $E_{\frac{1}{2}}$ for each analyte. |

SAQ 1.6a

SAQ 1.6b   Consult the data in Fig. 1.6c and comment on
                   the feasibility of analysing for Pb(II) and Tl(I)
                   ions in the same solution at 25 °C. What differ-
                   ence would it make if Pb(II) were monovalent
                   or if the concentrations of the two ions differed
                   greatly?

SAQ 1.6c

A series of standard solutions of $Zn(NO_3)_2$ were prepared and their polarograms were obtained separately at 25 °C. A solution containing Zn(II) ions at a concentration of approximately 10^{-4} mol dm^{-3} was now investigated (the unknown). The results are given below. Determine the accurate concentration of Zn(II) ions in the latter solution.

$10^4 \times c(Zn^{2+})/$ 0.50 2.00 3.00 5.00 unknown
mol dm^{-3}

$I_d/\mu A$.   0.65 2.58 3.86 6.46   4.75

Suppose the unknown had been a sample derived from a sewage effluent which contains many species other than Zn(II). Do you consider the method that you have just used to be satisfactory?

SAQ 1.6d

Why is dc polarography said to be a non-destructive method of analysis? Name one other non-destructive and one destructive method of analysis.

SUMMARY AND OBJECTIVES

Summary

A summary of the essential requirements for a dc polarography experiment is given in the form of a step-by-step description of the experiment. The resulting polarogram is discussed from the point of view of qualitative and quantitative analysis. The key to qualitative analysis is the value of the half-wave potential ($E_{\frac{1}{2}}$) and a method is described to give accurate values of this quantity. These values are seen to vary with solution conditions. Quantitative analysis is a matter of obtaining values of the diffusion current (I_d) which are linearly related to analyte concentrations. The direct calibration curve method and the standard addition method of obtaining a quantitative result are explained and compared. Dc polarography is categorised as a non-destructive method and as an example of a microelectrolysis method.

Dc polarography as an analytical method is not confined to the analysis of metallic cations.

Objectives

You should now be able to:

- describe a typical experiment designed to obtain the dc polarogram for a metallic cation, and obtain a value of the half-wave potential;

- explain the use of $E_{\frac{1}{2}}$ data as diagnostic for the reaction under investigation and discuss the possibilities for multi-cation analysis and state the criterion for resolution of neighbouring waves;

- explain the origin of a derivative polarogram and state its advantages over the normal polarogram;

- describe the calibration curve and standard additions method of obtaining a quantitative result from polarographic data and state the advantages of the latter method;

- perform calculations based on the calibration curve and standard additions method.

1.7. LIMITATIONS AND PROBLEMS

Since this closing section is devoted to a criticism of dc polarography we will start by reminding ourselves of the advantages of the DME, which is the feature that sets polarography apart from other voltammetric methods. These advantages are:

— the drops are reproducible in size and each drop grows in an
environment almost identical to that of its predecessor giving
rise to reproducible currents for the same conditions;

— a new surface is presented with each drop so that no appreciable
accumulation of reaction product at the surface can occur, either
as adsorbed species or as precipitates;

— species present in the solution, other than those involved in the
electrode reaction, are also unable to adsorb to any extent during
the lifetime of the drop;

— electrolysis is so small that solutions may be analysed many times
with no appreciable change in the analyte concentration;

— the overpotential for the reduction of the hydronium ion is high
so that interference from reduction of this species does not occur
until potentials more negative than -1.6 V (SCE) unless the pH
is low.

We can now turn to the limitations of the method and first of these
is the voltage limit of about 0 to -2 V (SCE) in aqueous solution.
This is not a serious limitation for metallic cation analysis but it is
for organic analysis. The use of non-aqueous solvents and supporting electrolytes such as tetraalkylammonium salts (1.4) enable the
negative limit to be extended to about -3 V (SCE). Little can be
done about the anodic limit which is determined by the tendency
of mercury to oxidise at potentials in the range 0–0.2 V (SCE).

The next and most important limitation is the high detection limit.
Dc polarography is best used for metallic cations in the range 10^{-2}–
10^{-4} mol dm^{-3}. At concentration $>10^{-2}$ mol dm^{-3} the large change
in concentration at the electrode surface causes the currents to become erratic. At concentrations $<10^{-4}$ mol dm^{-3} the capacitive
current (I_{cp}) becomes comparable with the diffusion controlled
Faradaic current. This sets a detection limit of at best 10^{-5} mol
dm^{-3}. In addition the depletion in analyte concentration near the
electrode during the initial stages of the growth of the mercury drop
causes a loss of sensitivity. It is generally accepted that experiments
done with unthermostatted cells are capable of giving results which

are reproducible to $\pm 2\%$. Provided great care is taken in the experiments and the cells are thermostatted then it is possible to achieve reproducibility of about $\pm 0.2\%$ when the analyte concentration is in the favourable range of $10^{-2}-10^{-4}$ mol dm^{-3}.

We have only touched upon complications or problems. We have mentioned the fact that some reactions are irreversible in the electrochemical sense and that although this changes the wave shape it does not necessarily prevent the use of the wave for analytical purposes. In passing we noted the effect of the nature of the supporting electrolyte on the half-wave potential (Fig. 1.6d). It is clear that the half-wave potential may be changed by changing the supporting electrolyte and this is used to selectively move waves when successive waves overlap. A good example is the effect of the presence of cyanide ions on the half-wave potential of Cd(II). The value of $E_{\frac{1}{2}}$ Cd (II) in 0.1 mol dm^{-3} KCl is -0.58 V (SCE), whereas in 1 mol dm^{-3} KCN it is -1.18 V (SCE). This is due to the formation of a complex cation with a wave that can still be used in the analysis for Cd(II). This is a matter of taking advantage of what is fundamentally a problem, ie any species capable of complexing with the analyte will alter the expected wave. In addition to these two problems there are three other effects which can cause problems.

— Catalytic hydrogen currents. These are only of importance in organic applications. They are caused by organic species adsorbed on the electrode that are capable of protonation.

— Adsorption currents. The reactant or product of the reaction adsorbs strongly or if the product is insoluble it simply adheres to the surface. This gives rise to an adsorption current.

— Kinetic currents. If the reactant or product is able to participate in a non-electrochemical reaction in the solution which either precedes, parallels or follows the electrode reaction then kinetic currents are caused.

All of these effects cause either distortion of the expected analyte wave or extra waves to appear, sometimes preceding and sometimes following the analyte wave. These effects will be pursued further in the next part of this Unit.

SAQ 1.7a   Summarise the main limitations and problems
               associated with dc polarography.

SUMMARY AND OBJECTIVES

Summary

The advantages of the use of the DME are listed.

Limitations include the restricted voltage window in dc polarography and the relatively high detection limit of about 10^{-5} mol dm^{-3}. The reproducibility of the method is routinely $\pm 2\%$ working at ambient temperatures and can be as good as $\pm 0.2\%$ when thermostatted cells are used for the method's optimum concentration range of 10^{-2}–10^{-4} mol dm^{-3}.

Problems can arise from irreversible waves, complexing, catalytic currents, adsorption currents and kinetic currents.

Objectives

You should now be able to:

- list the advantages of the use of the DME as a working electrode;

- discuss the limitations of dc polarography in terms of the voltage window, the detection limit and the reproducibility of the method;

- list the various effects that can cause problems in dc polarography.

2. Electrode Processes

Overview

You have already learnt that the height of a polarographic wave can be limited or controlled by the rate at which the analyte diffuses to the electrode. In this Part you will learn how to identify other types of polarographic waves and understand their role. These new types of polarographic current should generally be seen as potential interferences to be identified and avoided.

You will be introduced to the role of solvent, pH, and complexation in polarography, to permit an intelligent choice of a suitable supporting electrolyte for a given analytical problem. The use and advantages of electrodes other than the dropping mercury electrode will be considered.

2.1. Reversible and Irreversible Processes

Throughout this study you will meet the term thermodynamical reversibility. The degree of reversibility of the electrode reaction can have a profound effect on the polarographic behaviour and influence its analytical application. However for the present purposes a simple qualitative picture is sufficient.

Thermodynamical reversibility is not whether the reaction can run in reverse or not, but a question of energy and kinetics.

Cadmium ions (in KCl) undergo a thermodynamically reversible reduction to cadmium metal. However a certain percentage of the cadmium metal atoms formed are continuously reoxidised back to cadmium ions. Both the forward and this back reaction are rapid. The rate of the overall reaction from cadmium ions to the metal is a balance of the forward and back reaction rates with the forward dominating. It is the back reaction which determines whether the electrode process is thermodynamically reversible.

In a thermodynamically reversible electrode reaction the back reaction, from products back to the original species, occurs at an appreciable rate (at the potential of the wave) and must be taken into account. In a thermodynamically irreversible process the back reaction rate is negligible at the potential of the polarographic wave. In a reversible process the direction of the electrode reaction repeatedly reverses, for individual molecules or ions, between forward and back reactions. The overall reaction goes forward because the forward rate is the more rapid. An irreversible process is a one way only process and does not reverse at the potential of the wave.

Since a thermodynamically reversible electrode process is one in which the the overall reaction rate is a balance between the forward reaction to the product and the back reaction from product back to the original species, both the original species and the products must control the properties of the wave and the potential at which the overall reaction occurs. Thus the half wave potential of the wave, is controlled by the energy difference between the original species and the electrode reaction product. The half wave potential will lie close to the theoretical standard electrode potential for the reaction.

∏   What is the relationship between energy and potential?

A potential or voltage is itself a measure of energy available. Energy, for example, is often quantified in terms of electron volts, the energy taken to move one electron through a potential difference of one volt. More important a potential is a direct measure of the Gibbs free energy change in a reversible electrochemical reaction as expressed in the simple relationship:

$$\Delta G = - nFE$$

where ΔG is the free energy change per mole, and E the potential generated.

Iron(III) in some solutions produces a well formed cathodic (reduction) wave, in which the iron(III) undergoes a thermodynamically reversible reduction to iron(II). Since the potential of this wave is determined by the energy difference between iron(II) and (III), it could be expected that a solution of iron(II) in the same media would produce a well formed anodic (oxidation) wave at exactly the same potential. In both cases the potential of the wave would be determined by the same energy difference. In practice the two waves are separated by a very small difference in potential, ($59/n$ mV where n is the number of electrons transferred, in this case 1). The current in one wave flows in the opposite direction to the other. If both iron(II) and (III) are present in the same solution a single wave should develop with part cathodic and part anodic. The current direction will reverse part way up the wave. If the reversibility of the reaction is lost, through some addition to the solution, the cathodic and anodic waves will separate and move apart.

In a thermodynamically irreversible electrode process the rate of the back reaction, restoring the original species, is negligible. The potential of the reaction is concerned only with the original species and the forward reaction rate and not with the product. The potential could be thought of in terms of an activation energy required to get the reaction going.

| SAQ 2.1a | What is meant by the term activation energy? |

SAQ 2.1a

For some irreversible processes the back reaction is simply not possible. For example when an alkene is reduced to an alkane, it is often impossible to reoxidise the alkane back to the alkene at any available potential.

For other irreversible processes it is possible to reverse the electrode reaction but only at much increased potentials. Hydroxylamines undergo irreversible reduction to amines. At the potential of the reduction wave, the rate of the back reaction (reoxidation of the amine) is negligible. However at much more anodic (positive) potentials a solution of the amine can undergo oxidation. This oxidation is also an irreversible process and at these more anodic potentials re-reduction of the hydroxylamine back to amine has become negligible. Thus the cathodic wave (reduction of hydroxylamine to amine) and anodic wave (reoxidation of the amine) occur at widely separated potentials. At each wave the appropriate back reaction rate is negligible. In both cases the potential of the wave is determined by the initial species only and not by the electrode product. (The oxidation of the amine in fact usually occurs at potentials too anodic to be achieved at a dropping mercury electrode and requires a graphite electrode to be seen.)

The above can be summarised as follows:

— for a thermodynamically reversible reaction the reverse reaction from product to original species occurs at the same potential as the forward reaction, thus the respective polarographic half wave potentials or peak potentials should coincide at a potential determined by both original reactants and electrolysis products;

— thermodynamically irreversible electrode reactions either cannot be reversed or the reverse reaction takes place at a much higher potential. The potential of each irreversible process is determined by the original reactants only and not by the electrolysis products.

This will be discussed again in Section 2.7.

2.1.1.   The Potential Controlling Step

Many thermodynamically irreversible electrode reactions in fact take place in several steps. One of the steps, often the addition of the first electron, is the slowest step and thus this is the rate determining step. As the rate determining step it is also determines the potential of the reaction. We thus speak of the potential-determining step. Subsequent faster electrochemical reactions will not alter the half wave potential but merely add to the height of the wave.

Thermodynamically reversible electrode reactions can be effectively considered as occurring in a single step as only original and final products are important.

2.1.2.   Species Giving Rise to Several Waves

The above description applies to a single polarographic wave. A single species can, of course, give rise to several waves representing different overall processes. Each wave will have its own degree of reversibility.

Most nitrobenzenes, for example, give rise to two waves. The first wave is an overall four electron reduction to the hydroxylamine. The process is thermodynamically irreversible with the addition of

the first electron as the potential-determining step. The second wave
is due to a further two electron reduction of the hydroxylamine on
to the amine. This wave is also irreversible. The first is sharp and
has a well formed shape but the second wave is poorly formed (Fig.
2.1a.).

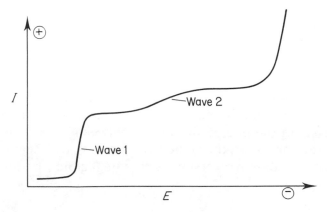

Fig. 2.1a. *The current-potential curve for a nitrobenzene*

$$PhNO_2 + 4e + 4H^+ \rightarrow PhNHOH + H_2O$$
$$PhNHOH + 2e + 2H^+ \rightarrow PhNH_2 + H_2O$$

Iron(III) in certain media gives rise to two waves, the first a one
electron wave with iron(II) as the product and second is due to the
reduction of this iron(II) to metallic iron. Both waves are thermodynamically reversible.

In other cases one wave may be reversible and the other not.

2.1.3. The Significance of Reversibilty

The significance of reversibilty in the analytical context lies in its
effect on the wave shape (or peak shape in some of the more advanced polarographic techniques). Irreversible processes give rise to
broader less steeply rising waves or broader peaks. While very many
irreversible processes do give rise to well formed waves suitable for
analytical application, with others the poor wave shape lowers their
precision and hence analytical usefulness. A good example would

be the nitrobenzenes mentioned above, the first wave (Fig. 2.1a) is suitable for polarographic analysis the second is not suitable. Those more advanced polarographic techniques which yield peaks rather than waves, show a loss in sensitivity and resolution as irreversibility makes the peak shorter but broader. This also renders the calibration useless.

The presence of surface active agents in the solution can sometimes lower the reversibility of the electrode reaction and thus alter the shape of the wave. The adsorbed material can create an additional energy barrier for the process to cross and reduce the reaction rates. If the process becomes dominated by a large activation energy barrier rather than the energy difference between the original species and the electrolysis product, then the process will cease to be reversible and become irreversible with a negligible rate for the back reaction.

2.2.　TYPES OF CURRENT FOUND IN POLAROGRAPHY

There are many types of current met with in polarography. Most types of current are not analytically useful. Fortunately the most useful, the diffusion controlled current, is also very common.

The polarographic signal usually consists of a wave. At low potentials no current flows. As the potential is raised a threshold voltage will be passed and the electrolytic process will begin. This threshold value is often known as the deposition potential. As the potential is raised further the current will rise exponentially. However it cannot continue to rise forever and eventually some process must limit the rise. Thus the rise in current will flatten out to form a second upper plateau, creating the typical polarographic wave. The current flowing at the upper plateau is known as the limiting current and is physically measured as the wave height.

A number of different processes can be responsible for limiting the rise in current. In essence, electrolysis can only occur as rapidly as the electro-active species is made available to the electrode surface. For example when the rise is limited by the maximum rate at which the analyte can diffuse to the electrode, we speak of a diffusion-controlled limiting current, or of a diffusion-controlled wave. Often the term diffusion current is used. When the kinetics or rate of a chemical reaction limits the electrolytic current, we speak of a kinetic current. And so on for other types of current.

Since some types of current are more useful or more reproducible than others it is necessary always to identify the type of current involved.

2.2.1. Faradaic Current

This is a general term for all currents caused by electrolysis of chemical species in the solution. It is so-called because electrolysis obeys Faraday's law. The term is used to differentiate electrolysis currents from currents, such as capacitive current, in which the cell is acting as an 'electronic' component. Diffusion current, kinetic current etc are all types of faradaic current.

SAQ 2.2a   What is Faraday's Law?

2.2.2. Diffusion Current

This is the only generally useful type of current from an analytical viewpoint. While other types of current occasionally find analytical application, diffusion current is generally the most reproducible type of current.

In a diffusion-controlled process the limiting current is controlled by the rate at which the species involved can diffuse to the surface of the electrode. As electrolysis continues the solution in the vicinity of the electrode becomes depleted of the electroactive species which must diffuse in from further out.

This would produce a drop in the current value, but the dropping mercury electrode itself is growing in surface area and growing out towards fresh solution. This would cause an increase in the current with time during the drop lifetime. These two opposing factors combine to give an overall rise in current during the drop lifetime.

By combining these two factors Ilkovic derived the famous equation bearing his name, already discussed in 1.5.2 (Eq. 1.5d). For a cathodic (reduction) process

$$I_d \ = \ 708 \, n \, D_{ox}^{\frac{1}{2}} \, m^{\frac{2}{3}} \, t^{\frac{1}{6}} c^{ox} \qquad\qquad (1.5d)$$

In classical dc polarography the average current during the drop lifetime is measured. For the average current I_d the Ilkovic equation becomes:

$$I_d \ = \ 607 \, n \, D_{ox}^{\frac{1}{2}} \, m^{\frac{2}{3}} \, t^{\frac{1}{6}} c^{ox} \qquad\qquad (1.5e)$$

Notice that the only difference in the two forms is the constant 607 or 708.

In these equations n and D are characteristic constants for the electroactive species and m and t are electrode parameters which can be kept constant. Therefore the limiting diffusion current is proportional to the concentration of the electroactive species in the solution. Thus the limiting diffusion current measured as the wave height can be used as an analytical tool to determine the concentration of the species.

Diagnostic criteria for diffusion currents are as follows.

— The wave height is proportional to concentration.

— With a freely dropping mercury electrode, the wave height is proportional to the square root of the height of the mercury column used. (Note that in order to observe this effect the drop must not be knocked off with the hammer attachment on most instruments. Allowance should be made for a small back pressure due to surface tension – about 1 cmHg.)

— If the current is measured during the drop lifetime it will be seen to rise in proportion to $t^{\frac{1}{6}}$. As the Ilkovic equation is only an approximation the diffusion current is usually proportional to $t^{0.19}$.

— Diffusion currents are relatively insensitive to temperature. The diffusion coefficient increases as the temperature rises, increasing the current by about 1.8% K^{-1}.

— Diffusion currents are usually independent of pH. With many organic compounds and some metal complexes the half wave potential, ie the position of the wave, is dependent on pH. The wave height however is usually independent of pH over wide ranges. The exception is in the vicinity of a pK_a value where the electroactive species will be undergoing protonation or deprotonation to another form. The wave height may change rapidly if the new form exhibits different polarographic behaviour.

∏   What is the difference between a proportional and a linear relationship?

When plotted as a graph both give a straight line. The proportional relationship ($y = mx$) passes through the origin (0,0) while the linear ($y = mx + c$) does not.

∏   The height of the mercury column does not appear in the Ilkovic equation. How can you explain that the current is dependent on the height of the mercury column, as stated above?

     The Ilkovic equation contains terms for the mercury flow rate and the drop lifetime. Increasing the height of the mercury column will clearly increase the flow rate and shorten the drop lifetime. If a freely dropping electrode is used, m is proportional to h and t is proportional to h^{-1} (see Section 1.3.2). The arithmetic details are not important to you, the result as a diagnostic tool is.

Diffusion in an unstirred solution is a highly reproducible phenomenon. Since the height of a diffusion-controlled wave is proportional to concentration, and is independent of pH and relatively insensitive to temperature, it is highly suited to analytical application.

2.2.3. Migration Current

This current is caused by the acceleration of ions towards (or away from) the electrode by electrostatic attraction. It is not a useful type of current. It does not have a linear relationship with concentration. It is usually supressed by adding at least a fiftyfold excess of an inactive supporting electrolyte which shields the ions of interest from the electrostatic attraction.

SAQ 2.2b   What effect would migration current have on the overall limiting current, or wave height, for (*i*) electrostatic attraction and (*ii*) repulsion, respectively, of the analyte to the electrode?

SAQ 2.2b

2.2.4. Kinetic Currents

There are many types of kinetic current. In all of these the electroactive species is not originally present in the solution, but is formed by a preceding chemical reaction in the solution or sometimes on the electrode surface.

The rate of the chemical reaction is crucial. An extremely slow reaction would show up as a slowly rising conventional diffusion-controlled wave. However if the rate of chemical production is faster than the rate of diffusion, the limiting current will be diffusion controlled. For a kinetic wave to form, the rate of the chemical reaction must be relatively rapid but significantly slower than the rate of diffusion.

A good example would be the reduction of formaldehyde (methanal). In aqueous solution formaldehyde exists as its monohydrate $CH_2(OH)_2$. This is not electroactive and must undergo dehydration to the free aldehyde before it can undergo reduction to methanol.

$$CH_2(OH)_2 \xrightarrow[-H_2O]{k_f} HCHO \xrightarrow{2e, 2H^+} CH_3OH.$$

The limiting current is almost entirely controlled by the rate of the dehydration reaction.

For such a simple first order reaction the kinetic current I_k is given by

$$I_k = 528nm^{\frac{2}{3}}t^{\frac{2}{3}} D_{ox}^{\frac{1}{2}} k_f^{\frac{1}{2}} K^{\frac{1}{2}} c^{ox} \qquad (2.2a)$$

where k_f is the forward rate constant (s^{-1}) of the preceding chemical reaction and K is the equilibrium constant of the reaction.

Diagnostic criteria for kinetic currents are as follows.

— The wave height is proportional to concentration.

— With a freely dropping mercury electrode, the wave height is independent of the height of the mercury column used.

— If the current is measured during the drop lifetime it will be seen to rise in proportion to $t^{\frac{2}{3}}$.

— Kinetic currents are highly sensitive to changes in temperature, the current increases with rising temperature. This is because the rates of chemical reactions are themselves highly dependent on temperature. As the temperature rises the kinetic current will rise dramatically, often in the range of 10–20% K^{-1}. If the temperature rises high enough the rate of the chemical reaction will exceed the rate of diffusion, and the wave will become diffusion-controlled.

— Kinetic currents are often highly sensitive to pH changes, because rates of chemical reactions are often, but not always, sensitive to pH.

— Kinetic currents are generally smaller than the corresponding diffusion current.

The above equation (2.2a) and diagonistic criteria only apply strictly to simple first order reactions preceding the electrolysis. The details may vary for more complicated kinetics.

∏ How might the relationship between the current and time be experimentally measured during the drop lifetime which lasts only 1 or 2 s?

An oscilloscope could be used, with a memory or a photographic record. Most chart recorders would not have a fast enough reaction time to allow accuracy.

SAQ 2.2c   From experimental data how might one determine to which power of the time the current is proportional?

With a proportional relationship between the current and the concentration, kinetic waves might seem suitable for analytical application. However, because kinetic currents are so sensitive to temperature and pH, sufficient reproducibilty and accuracy can rarely be obtained, even with the best temperature control generally available.

2.2.5. Diffusion Currents with a Kinetic Contribution

This also involves a chemical reaction preceding the electrochemical step. In this case the rate of the preceding chemical reaction is more similar to the rate of diffusion and the limiting factor on the current is a combination of both diffusion control and the rate of the chemical reaction.

A typical example would be the reduction of some organic acids, such as fumaric acid, at pH values in the region of their pK values. Both the acid and its anion will be present in the solution in an equilibrium ratio. If only the acid form is electroactive at the given potential, we might expect the height of the wave to correspond to the equilibrium concentration of the acid form in a diffusion-controlled wave. However by removing the acid form by electrolysis, we have disturbed the equilibrium. The equilbrium will attempt to restore itself by protonation of the anion. This chemical reaction will produce more electroactive acid and the limiting current will be increased. The increase is known as the kinetic contribution since it depends on the rate at which the protonation reaction occurs. Well away from the pK value this rate will be negligible or the compound will be totally in the electroactive form.

Diagnostic criteria for diffusion currents with a kinetic contribution are as follows.

— The wave height is proportional to concentration.

— With a freely dropping mercury electrode, the wave height is proportional to somewhere between h^0 and $h^{\frac{1}{2}}$ where h is the height of the mercury column.

— If the current is measured during the drop lifetime it will be seen to rise in proportion to a power of t between 2/3 and 1/6.

— These currents are highly sensitive to changes in temperature, the current increasing with rising temperature. As the temperature rises the kinetic component will rise dramatically. If the temperature rises high enough the rate of the chemical reaction will exceed the rate of diffusion, and the wave will become totally diffusion-controlled and dependent on the sum of both electroactive and electroinactive forms.

— The kinetic component of these currents is often highly sensitive to pH changes, because rates of chemical reactions are often, but not always, sensitive to pH.

Again this type of current is best avoided for analytical purposes. Usually it is possible by choice of temperature, pH etc to find conditions where the kinetic component is negligible and the wave is purely diffusion controlled.

2.2.6. Catalytic Currents

These are also kinetic currents except, that now the chemical reaction or the electrolysis regenerates the original substance on which the height and position of the wave depends. Since this substance is not consumed by the electrolysis, it plays the part of a catalyst and hence the wave it generates is known as a catalytic wave.

The most common variety is the catalytic hydrogen wave. A good example is the catalytic wave generated by pyridine. Pyridine undergoes a protonation reaction to the pyridinium ion.

$$C_5H_5N + H^+ \longrightarrow C_5H_5NH^+$$

$$C_5H_5NH^+ + e \longrightarrow C_5H_5N + 0.5 H_2$$

The pyridinium ion is then reduced with the regeneration of the pyridine and hydrogen gas liberated. In practice the overall reaction is the reduction of the hydrogen ion to hydrogen gas. However this occurs at a potential significantly lower than that for the normal reduction of the hydrogen ion and forms an independent wave whose height is controlled by the concentration of the pyridine. Because the pyridine is not consumed and is rapidly regenerated, the limiting current is very much larger than that of a diffusion-controlled reduction of pyridine.

Unfortunately catalytic currents are usually of poor reproducibility and thus of limited analytical use. On the other hand being often two orders of magnitude larger than diffusion currents they do offer greater sensitivity. An analytically useful example is the determination of molybdate by means of its catalytic reduction of nitrate. The molybdate ion, Mo(VI), is reduced at the electrode to molybdenum(III) ions. In an excess of nitrate the molybdenum(III) ion reduces the nitrate ion and is itself reoxidised to molybdenum(VI) in the subsequent chemical step. The regenerated molybdenum(VI) is available for more reduction and the subsequent catalytic wave is about 200 times greater than the diffusion controlled wave in the absence of nitrate. In the presence of a large excess of nitrate the wave height depends only on the original molybdate concentration.

Diffusion currents can also be increased by a catalytic component. This is usually a nuisance. For example in some supporting electrolytes chromium(III) produces a one electron wave in which it is reduced to chromium(II). However chromium(II) is a rather unstable and powerful reducing agent, and some of it is reoxidised by water or other solution components back to chromium(III). This regeneration adds a catalytic component to the diffusion-controlled wave. This addition to the wave height is of poor reproducibility and renders the wave unsuitable for analytical application. Chromium(III) is best determined polarographically by using thiocyanate as a complexing agent in the supporting electrolyte. The thiocyanate complex of chromium(III) exhibits a three electron wave, being reduced to chromium metal.

∏   Write down the relevant chemical equations for the various steps of the catalytic component of a wave in which some of the chromium(II), produced in a one electron reduction of chromium(III), reduces water. Write down the overall electrode reaction for this catalytic component.

$$Cr^{3+} + e \longrightarrow Cr^{2+}$$

$$Cr^{2+} + H_2O \longrightarrow Cr^{3+} + 0.5H_2 + OH^-$$

$$H_2O + e \longrightarrow 0.5H_2 + OH^-$$

In acid solution this becomes:

$$Cr^{3+} + e \longrightarrow Cr^{2+}$$

$$Cr^{2+} + H^+ \longrightarrow Cr^{3+} + 0.5H_2$$

$$H^+ + e \longrightarrow 0.5H_2$$

In practice this reaction with water is relatively slow and most of the electrode reaction is the reduction to Cr(II) which diffuses away into the solution bulk. The catalytic component would be relatively small. It still represents a significant uncertainty and other unknown solution components could produce a much larger catalytic component. This wave due to reduction of chromium(III) to chromium(II) is quite unsuitable for analytical application. Conditions in which the chromium(III) is reduced to the metal in one step should be sought.

Diagnostic criteria for catalytic currents are as follows.

— The wave height is often proportional to concentration, but in other cases may tend towards a limiting maximum height. This will often depend on whether the catalytic reaction occurs in the solution or on the surface of the electrode.

∏   If a catalytic reaction occurs on the surface of the electrode why should the wave height tend to a constant maximum with increased concentration?

The surface of the electrode contains only a fixed number of sites on which the species concerned can adsorb. The number of surface sites is limited and once fully occupied, increasing the concentration will have little effect.

— With a freely dropping mercury electrode, the wave height is frequently independent of the height of the mercury column. But many different dependencies on the height of the mercury column are found varying with the nature of the catalysis. A wave height which increases as the height of the mercury column decreases will always be a catalytic process.

— If the current is measured during the drop lifetime it will typically be seen to be proportional to $t^{\frac{2}{3}}$, although other dependences are also known.

— Catalytic currents are highly sensitive to changes in temperature, because the rates of the chemical reactions involved are themselves highly dependent on temperature.

— Catalytic currents are often highly sensitive to pH changes, because rates of the chemical reactions are often, but not always, sensitive to pH. Catalytic hydrogen waves are particularly sensitive to pH decreasing very rapidly with increasing pH.

— Catalytic currents are generally very much larger than the corresponding diffusion current. Kinetic currents, in contrast are smaller than the corresponding diffusion current.

The above criteria present a confused picture because of the varied nature of catalytic processes. In general there is a resemblence to kinetic currents. The most telling diagnostic criterion is that catalytic waves are much larger than diffusion-controlled waves.

Catalytic waves are generally not suitable for analytical application, as most are not very reproducible. When a catalytic wave is applied, for its greater sensitivity, great care must be taken with the conditions pH, temperature, etc to ensure an acceptable level of reproducibility. Diffusion-controlled waves with catalytic components are best avoided altogether.

| SAQ 2.2d | What is the essential difference between a catalytically and a kinetically controlled wave? |

2.2.7. Adsorption Waves and Phenomena (A Polarographic Chamber of Horrors)

Adsorption phenomena are the greatest single hazard facing the analyst using polarographic methods. They come in many varied forms, most of them as interferences in analytical applications. Most are highly irreproducible. The analytical chemist using polarography must always consider the danger of surface active agents being present in his solutions and samples.

The adsorbed species causing any given adsorption effect can either be the electroactive species of interest or some other species in the solution. One species can itself give rise to several different adsorption phenomena, as one chemical species is capable of existing in several different adsorbed forms.

∏   What is the difference between adsorption and absorption?

Adsorption is a surface phenomenon. Absorption is a bulk phenomenon. They must not be confused.

∏   How can one chemical species have several different adsorbed forms?

Each form represents a different orientation of the molecule to the adsorbing surface, or differences in the bonding to the surface and the bonding group. The more complex the molecule the more potential adsorbed forms are likely to be found.

(a) Adsorption prewaves and post waves.

This is the simplest and most reproducible type of adsorption phenomenon. With many polarographic waves, the main diffusion-controlled wave is accompanied by a small preceding or following wave of constant height (Fig. 2.2a). This is an adsorption prewave or postwave, sometimes known as a Brdicka adsorption prewave or postwave.

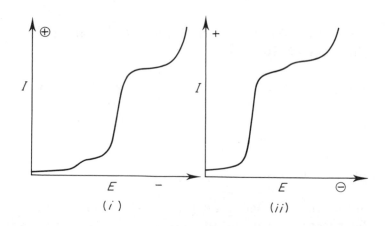

Fig. 2.2a.   *A diffusion-controlled wave with an adsorption prewave (i) or postwave (ii)*

The prewave is caused by the adsorption on the surface of the electrode of some of the electroactive material. If the adsorbed form is easier to reduce then it will be reduced at a lower potential than the rest of the electroactive form in the solution. The adsorbed form will thus form a separate prewave at more positive potentials (for a reduction or cathodic process) than the main wave. Since there are only a limited number of sites available for adsorption, they will become fully occupied above a certain concentration. Thus the height of the adsorption prewave will become constant after a certain concentration and any further increase in concentration will only increase the height of the main wave.

The current flowing in the adsorption prewave has in fact been taken from the main wave as the adsorption took place. It is the sum of the current in both main and prewave which is proportional to the concentration of the electroactive species. The sum is diffusion-controlled.

Adsorption postwaves are similar to the prewaves but are caused by adsorption of the electrolysis product, or when the adsorption of the original electroactive form makes it more difficult to reduce. Usually adsorption of a species releases energy which is then available to the electrolysis step, making it easier to reduce the species.

Diagnostic criteria for adsorption currents are as follows.

— The wave height is independent of concentration, above a certain concentration.

— With a freely dropping mercury electrode, the wave height is proportional to the height of the mercury column.

— If the current is measured during the drop lifetime it will be seen to fall in proportion to $t^{-\frac{1}{3}}$.

— Adsorption pre- or post-waves are usually independent of changes in temperature. Occasionally an increase in temperature can cause a fall in the adsorption current as the species desorb.

— Adsorption pre- and post-waves are usually independent of pH within wide ranges. Although crossing a pK value may remove the adsorbing species from the solution.

— Addition of a more strongly surface active agent which is itself electroinactive (eg Triton X-100) can oust the adsorbed form from the solution destroying the pre-or post-wave. The current then returns to the main wave.

Pre- and post-waves of this kind can often be tolerated, provided the sum of the currents for the prewave and main wave is taken as the analytical signal. They can be avoided at times by choosing a different solvent or supporting electrolyte. The use of a surfactant to remove prewaves is usually best avoided.

(*b*) Polarographic Maxima.

This phenomenon has been briefly discussed in 1.3.3. Sometimes polarographic waves have large maxima. There are two kinds. 'Maxima of the first kind' are found as continuations of the rising part of the wave and are usually sharp (Fig. 1.3f). 'Maxima of the second kind' are rounded humps found on the upper plateau of the wave. Both are adsorption phenomena. Maxima of the first kind are believed to be caused by rapid streaming of the solution past the electrode, somehow caused by adsorbed forms. Maxima of the second kind involves streaming of the mercury within the mercury drop.

∏   Why should this streaming increase the current?

The streaming of solution past the electrode will increase the amount of fresh analyte reaching the electrode and so the current will increase. In maxima of the first kind this only occurs on the rising part of the wave and stops soon after and so the current collapses back to the normal limiting value, thereby forming the sharp maxima. Exactly why adsorption effects should produce this streaming is not clear.

Maxima can usually be eliminated by adding an electroinactive surfactant such as Triton X-100.

(*c*) Inhibition and related phenomena.

Adsorbed species can cause energy barriers to the progress of the desired electrolysis. This can produce a wide variety of phenomena. The shape of the wave may change, frequently becoming less steep and often poorly formed. The wave may be moved to higher potentials. The wave height can become suppressed, sometimes the wave disappears altogether. At other times the wave contains sharp discontinuities as the degree of inhibition changes dramatically as one adsorbed form rearranges to another or is desorbed with the change in potential.

Fig. 2.2b shows an extreme example of the effects inhibition can have. The compound is triphenylarsine oxide, which is reduced in a two electron step to the arsine. The many adsorbed species causing these inhibition effects are the many adsorbed states of triphenylarsine oxide itself. A simple wave is formed when these adsorbed forms are ousted from the surface of the electrode by adding a very small quantity (eg 0.005% v/v) of the more powerfully adsorbed Triton X-100 (a polyalcohol). The latter, although powerfully adsorbed, has only relatively small inhibition effects on many electrochemical processes.

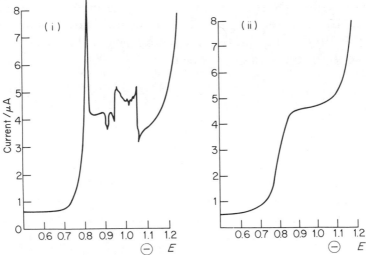

Fig. 2.2b.   *The current/potential curve for 0.001 mol dm^{-3} triphenylarsine oxide in 0.1 mol dm^{-3} HCl (i) without and (ii) with the addition of 0.005% (v/v) Triton X-100*

Fortunately the more bizarre effects are only met with in the case of the some of the more strongly surface active species. The most commonly found effect of surfactants is a shift in potential and a suppression of the wave height. The latter would of course alter the calibration plot of the method. The analytical chemist should always keep his eye open for the possibility of inhibition effects altering the precision of his work.

A change of solvent, for example the addition of an organic solvent to an aqueous solution, will frequently alter or remove interfering inhibition effects. The use of a more harmless but stronger surfactant to oust the interfering adsorbed forms is also possible but must be applied with great caution.

2.2.8. Capacitive Current

This topic has been dealt with in some detail in 1.3.3 but you are reminded of some essential points. Capacitive current is not a faradaic current, that is, it is not due to an electrochemical reaction. Rather it is due to the electrode/solution interface acting as a electrical component, to be exact as a capacitor. When a negative voltage is applied to the electrode, the surface of the mercury metal will be covered by negative charges. These in turn will attract a layer of positive ions from the solution. These two layers (Fig. 2.2c) of charge act like the two plates of a capacitor.

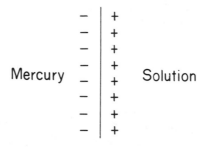

Fig. 2.2c.   *The Helmholtz double layer at the electrode surface*

The voltage between the plates, across the electrode/solution interface, (that is the potential of the electrode) depends directly on the charge density on the mercury surface. To impose a potential on the electrode a current must flow in order to create the charges on the electrode surface. This is the capacitive current.

If a constant potential is imposed on a solid electrode of fixed size this capacitive current dies off rapidly once all the charges are in place. However the dropping mercury electrode is growing in surface area all the time, fresh charge is required just to keep the charge density constant. Thus a capacitive current continues to flow.

This capacitive current does not usually form discrete waves of its own but is the major cause of the residual current. This capacitive current increases with increasing potential and so creates the slope of the upper and lower plateaux of polarographic waves. It is a background signal, a source of noise, and as such it is a major limiting factor on the sensitivity of dc polarography.

An electrical capacitor contains a dielectric medium between the two plates. The higher the dielectric constant, the more charge is needed on the plates to maintain the voltage difference. Species adsorbed on the mercury electrode can act as a dielectric medium. Thus the capacitive current varies somewhat with adsorption on the electrode and is not predictable from electrode charactaristics alone.

The decrease of the capacitive current with time during the drop lifetime is proportional to $t^{-\frac{1}{3}}$ (Fig. 1.3d).

| SAQ 2.2e | The height of a dc polarographic wave was measured at different heights of the mercury column for a freely dropping mercury electrode. The following wave heights were obtained, for these mercury column heights corrected for back pressure. $\longrightarrow$ |

SAQ 2.2e
(cont.)

height of column/cm   30   40   50   60   70

wave height/arbitrary
  units   23.5 27.3 32.0 33.5 36.2

A good proportional relationship was found for the wave height with concentration. The wave height was found to increase with temperature by about 2% K^{-1}. Which type of limiting current was involved?

SAQ 2.2f

A substance of interest gave rise to a wave whose height was independent of the height of the mercury column. The wave height was proportional to the concentration and increased with temperature by about 15% K^{-1}. Which type of limiting current was involved?

SAQ 2.2f

SAQ 2.2g

The dc polarographic wave height was found to be proportional to the concentration of the analyte. The wave height was further found to be proportional to the square root of the height of mercury column while the current during the drop lifetime was found to be proportional to $t^{0.23}$, where t is the time from the birth of the drop. The wave height increased by 12% K^{-1} with rising temperature. Which type of limiting current was involved?

SAQ 2.2h

The height of a dc polarographic wave was measured at different heights of the mercury column for a freely dropping mercury electrode. The following wave heights were obtained, for these mercury column heights corrected for back pressure.

height of column/cm   30   40   50   60   70
wave height/arbitrary
    units   15.7 17.3 18.0 19.2 20.0

A reasonably proportional relationship was found for the wave height with concentration. The wave height was found to increase with temperature by about 7% K^{-1}. Which type of limiting current was involved?

SAQ 2.2i

> The main polarographic wave of interest was preceded by a small prewave. The height of this prewave was independent of the concentration of the analyte. The wave height of the prewave was found to be proportional to the square root of the height of mercury column while the current during the drop lifetime was found to be proportional to $t^{0.23}$, where t is the time from birth of the drop. The height of the prewave increased only to a small extent with rising temperature. What could be the origin of this prewave?

SAQ 2.2j   Which are the only types of current to decrease during the drop lifetime?

SAQ 2.2k   Why is it best to avoid the region around pK values for polarographic determination of an analyte?

2.3. THE ROLE OF pH

Hydrogen ions are involved in the reduction of most organic species. Since the reduction often involves adding electrons to a neutral molecule to form a reduced species, which itself is a neutral molecule, then an equal number of hydrogen ions are required to ensure neutrality.

$$\text{ox} + n\text{e} + n\text{H}^+ \longrightarrow \text{red}$$

Since the hydrogen ions are involved, the hydrogen ion concentration must be included in the Nernst equation for the reaction.

$$E = E^{\ominus} + \frac{RT}{nF} \ln \frac{c^{\text{ox}} c_{\text{H}^+}^n}{c^{\text{red}}}$$

(strictly speaking activity terms and not concentrations should be used, but in the dilute solution used there is usually little difference).

It can therefore be seen that the potential is dependent on the hydrogen ion concentration, that is on the pH of the solution. The Nernst equation applies only to a thermodynamically reversible system but an equivalent expression involving the hydrogen ion concentration applies to all reductions or oxidations.

The half wave potential of the polarographic wave for such a reaction will also be dependent on pH. The reaction above shows an equal number of hydrogen ions consumed as electrons. For the more general case involving n electrons and p hydrogen ions the half wave potential will be:

$$E_{\frac{1}{2}} = E_{\text{const}} + \frac{RT}{nF} \ln c_{\text{H}^+}^p$$

or

$$E_{\frac{1}{2}} = E_{\text{const}} - \frac{pRT}{nF} \text{pH}$$

For a thermodynamicaly irreversible process the equation must be modified by the addition of a constant α with a value between 0 and 1.

$$E_{\frac{1}{2}} = E_{\text{const}} - \frac{pRT}{\alpha nF}\, \text{pH}$$

Thus it can be seen that the half wave potential for the reduction or oxidation of a species in the solution shifts at a constant rate per pH unit, dependent on the number of hydrogen ions and electrons in the potential determining step. For a reduction the half wave potential moves to more negative potentials with increasing pH, ie. the reduction process becomes more difficult. For an oxidation the half wave potential also moves to more negative potentials, ie. the oxidation process becomes easier.

Many electrochemical reductions or oxidations are several step procedures. Often the first of these is the rate-determining step and hence the potential-determining step. Subsequent steps are much more rapid and do not effect the potential but merely the height of the wave. The shift of the half wave potential with pH is determined by the number of hydrogen ions involved in and before the potential-determining step.

Π   How could the reversibility of the potential-determining step be obtained from a study of the effect of pH on the polarographic behaviour of a compound?

Away from pK_a values the slope of the plot of the half wave potential against pH is given by $2.303pRT/nF$ for reversible processes and $2.303pRT/\alpha nF$ for irreversible processes, where p and n are the number of protons and electrons involved. If we substitute the universal constant values we get $0.059p/n$ V for reversible processes and $0.059p/\alpha n$ V for irreversible processes. That is simple integer multiples of 59 mV for reversible processes and non-integer for irreversible. Note that this applies only to the potential-determining step.

It is most important that the solution be adequately buffered if hydrogen ions are involved in the electrolysis. For a reduction in an unbuffered solution the electrolysis will consume hydrogen ions, lowering their concentration and raising the pH at the electrode surface. This in turn will shift the potential of the reaction. The result would be a long drawn out and very poorly formed wave. However if the solution contains a pH buffer at a high concentration (>0.01 mol dm^{-3}) the pH can be kept constant effectively down to the electrode surface. This will allow a well formed wave to be produced.

The reduction of metal ions does not usually involve hydrogen ions and thus well formed waves can be obtained in unbuffered solutions. However the stability of many complexes formed between the metal and ligands in the solution is dependent on pH. Since changes in complexation will alter the polarographic behaviour of the metal ion, it is best to use a well buffered solution if complexing agents are present.

2.3.1. Changes of Polarographic Behaviour with pH

The above discussion describes the shift in half wave potential with pH for an individual species in the solution. However as the pH changes so the nature of the species in the solution will also change as acid/base reactions take place and the analyte is found in different states of protonation.

The different protonated forms of an individual compound can vary considerably in their polarographic behaviour. Often only one protonated form is electroactive. In other cases a second protonated form is also electroactive but at a quite different potential. In some cases the electroinactive form can be rapidly protonated or deprotonated to the electroactive form in a fast step before the electrolytic step, giving rise to the wave but at a shifted potential. In other cases the electroinactive form remains totally inactive and the wave vanishes when the pH is adjusted to a value where this form is dominant. Sometimes the mechanism of the reduction or oxidation can change with pH.

The role of pH in polarography can be very varied. An entire textbook would be needed to cover all the many possibilities. The best that we can do in our present study is to look at a few examples to illustrate the range of common behaviour. For detailed information a textbook such as Zuman (1967) should be consulted.

The antihistiminic drug triprolidine consists of a substituted olefin and undergoes a two electron reduction of the double bond in a single well formed wave.

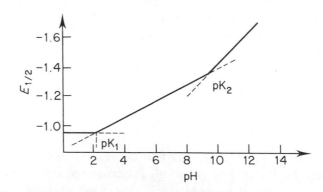

Triprolidine contains two nitrogen atoms capable of protonation and therefore exists in three states of protonation with two pK_a values. The electroactive form is the doubly protonated dicationic form. However the monoprotonated and the unprotonated forms are both capable of undergoing very rapid protonation near the electrode surface. As a result, the wave occurs throughout the pH region and the height is independent of pH. When the monoprotonated or unprotonated forms are present the necessary protonation prior to the electroreduction simply adds to the number of hydrogen ions consumed in the potential-determining step. As a result the shift of half wave potential with pH becomes steeper.

Fig. 2.3a.   *The half wave potential of triprolidine as a function of pH*

A plot (Fig. 2.3a) of the half wave potential against pH consists of three intersecting straight lines, whose slopes correspond to 0, 1 and 2 hydrogen ions respectively involved in the potential-determining step. The intersection of the lines occur at the two pK_a values. At pH values below the pK_1 value the doubly protonated form is present in the solution and no hydrogen ions are consumed in the potential-determining step. Above the pK_1 value the monoprotonated form is present and 1 hydrogen ion must be added prior to the electrochemical step. Above the pK_2 two protons are required. Polarographic analysis of triprolidine is possible at any pH value but probably is best carried out between pH 1.7 and 6.0 for the maximum precision to be obtained.

∏　　How might one determine the best pH for the determination of a compound such as triprolidine which is electroactive throughout the pH range?

　　Simply by experimental trial. Ultimately there is no substitute. The complete method would be tried with representative standard samples at a range of pH values. A compomise between precision, accuracy and freedom from interferences would be chosen. The choice would not be made from theory except to eliminate the obvious.

A complete contrast is given by the phenylarsonic acids and diphenylarsinic acids. (The former are used as veterinary antibiotics and the latter are found as byproduct impurities in their manufacture.)

Phenylarsonic acid　　　　Diphenylarsinic acid

The anionic forms of these acids are completely electroinactive and do not protonate rapidly enough to form a polarographic wave. Only the free acidic form is electroactive. Both the arsonic and arsinic acids give rise to a single wave whose height is independent of pH at values well below the pK_a value. However as the pK_a is approached the wave height falls rapidly until the wave disappears completely (Fig. 2.3b).

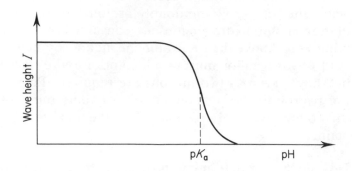

Fig. 2.3b   *The dependence of the wave height on pH for diphenylarsinic acid*

At higher pH values these compounds are no longer electroactive. As the pK_a value for the arsinic acids is about 7 the compound is electroactive below this value. The pK_a for the arsonic acids is about 3 to 4 and thus they are electroactive over a narrower range than are the arsinic acids. At a pH value of 1.5 the arsonic and arsinic acids produce waves of similar half wave potentials and so cannot be differentiated. However at pH 5.5 only the arsinic acids are electroactive. Thus the control and choice of pH allows the specific analysis of both the arsonic acids and the arsinic acid impurities.

∏   Why might the individual pH values 1.5 and 5.5 be chosen for the determination of the arsonic and arsinic acids?

It is best to get as far away from the pK_a values as possible to avoid interference and kinetic contributions. pH 5.5 is about half way between the pK_a values of the arsonic and arsinic acids, at pH 3–4 and pH 6–7. Similarly pH 1.5 is well away from the pK_a of the arsonic acids. However the actual choice should be made by experimental trial.

With the arsonic and arsinic acids the anionic forms are not electroactive. When both acidic and conjugate basic forms of a substance are electroactive but at different potentials, two separate waves will be observed near the pK_a value. As the pH is increased the wave corresponding to the basic form will increase at the expense of the wave due to the acidic form. A single wave would be found away from the pK_a value. However the height of the two waves does not generally reflect the actual concentration of the two forms as the waves frequently display kinetic components.

Frequently the mechanism of the electrolysis changes with pH. For example, secondary nitrosoamines undergo a 4 electron reduction in acid solutions to the unsymmetrical dialkylhydrazine.

$$R_2NN = O + 4e + 4H^+ \longrightarrow R_2NNH_2 + H_2O$$

but in neutral and alkaline solutions the wave falls to half its height as the mechanism changes to a 2 electron reduction to the secondary amine and nitrous oxide.

$$R_2NN = O + 2e + 1.5H_2O \longrightarrow R_2NH + 0.5N_2O + 2OH^-$$

With many substances the change of mechanism with pH can take the form of appearance or disappearance of waves at different pH values. Some multistep reductions or oxidations can occur as a single wave at one pH value but split into separate waves in other pH regions.

2.3.2. pH and Polarographic Analysis of Organic Compounds

The relationship between pH and the polarographic behaviour of organic compounds can be very complicated, and is in many ways the concern of the physical organic chemist rather than the analytical chemist. You should not need to remember the details of the previous section but merely use it to get a feel for the possibilities that might be involved. For the analyst several considerations are important and must always be considered.

Standard and sample solutions must always be buffered at the same pH value.

For the polarographic analysis of an organic compound a pH should be chosen where the wave height is independent of pH and is diffusion-controlled.

SAQ 2.3a	If the wave height is independent of pH, why should it be necessary to buffer samples and standards at the same pH?

The regions around pK values should always be avoided if possible. The wave height may be pH-dependent near the pK value. Worse still, protonation or deprotonation reactions prior to the electrolysis often introduce some kinetic character to the wave in the region of the pK value. The kinetic component renders the wave very temperature-dependent and leads to a loss in reproducibility.

The choice of pH can be used to enhance selectivity. A pH region may be chosen so that interfering substances are no longer electroactive. In the example (2.3.1) phenylarsonic acid would interfere in the determination of diphenylarsinic acid at pH 1.5 but by carrying out the polarographic analysis at pH 5.5 the phenylarsonic acid will cease to be electroactive and will no longer interfere.

Two overlapping waves interfering with each other can often be resolved by adjusting the pH. The two waves may well shift with pH at different rates. By adjusting the pH in a suitable direction the waves will move apart until they can be resolved satisfactorily. For example the half wave potential of one wave may be pH-independent while that of the other may be pH-dependent.

A further consideration is the decomposition of the supporting electrolyte which limits the useful working potential range. A pH must be chosen to move the wave of interest so that it will not be overlapped or partially obscured by the final decomposition of the supporting electrolyte.

SAQ 2.3b	Would you expect the potential of the reduction or oxidation of the supporting electrolyte itself to shift with pH?

The pH of the solution may also affect the stability of the analyte.

A good example of these last two considerations is the anodic determination of vitamin C (ascorbic acid) in an acetate buffered solution.

Ascorbic acid   Dehydroascorbic acid

Ascorbic acid is rather unstable with respect to oxidation and the solution should be as acidic as possible to promote stability. However if the solution is too acidic the polarographic wave due to the ascorbic acid moves into the region of decomposition of the supporting electrolyte and becomes obscured. A compromise pH should be sought.

2.4.   THE ROLE OF SOLVENT

Water is the most common solvent system used for polarography but many other solvents such as dimethylformamide (DMF), acetonitrile, dimethyl sulphoxide (DMSO) are also widely used. In general polar solvents must be used as they must be able to dissolve a suitable ionic salt, such as tetraalkylammonium salts, to act as a supporting electrolyte. The effect of the solvent is often profound as it greatly affects the stability of electrolytic intermediates. The biggest contrast is between protic solvents such as water and the alcohols which readily supply hydrogen ions and the aprotic solvents which do not.

A good example would be the reduction of diphenylalkenes. In aqueous solutions the diphenylalkenes are reduced in a single two electron wave to the diphenylalkane. In thoroughly dried acetonitrile the diphenylalkenes give rise to two separate well defined one electron waves.

$$Ph_2C=CH_2 + e + H^+ \longrightarrow Ph_2\dot{C}-CH_3$$

$$Ph_2\dot{C}-CH_3 + e + H^+ \longrightarrow Ph_2CH-CH_3$$

In aqueous solutions the radical produced in the first step is highly unstable and reaction proceeds immediately to the final product, forming a single two electron wave. In dry acetonitrile the radical is sufficiently stable to escape the electrode surface and eventually dimerise, thus forming a one electron wave. The second step in dry acetonitrile requires a higher potential and so the final product is only reached in a second one electron wave.

The addition of only a very small quantity of water renders the intermediate radical unstable and the two waves coalesce. Frequently polarographic behaviour in a completely dry aprotic solvent is quite different to the same solvent slightly 'wet' with traces of water.

∏   Would you describe an AnalaR solvent in an unopened bottle as dry or slightly 'wet'?

Very definitely wet in this context. Commercially available solvents must always be carefully dried and carefully stored, for example over molecular sieves, in order to count as very dry solvents. Very dry solvents are very unstable with respect to their dryness.

The maintenance of very dry solvents is rather difficult and does not lend itself to analytical applications where very high reproducibilities are required. Thus the use of these solvents is largely in the field of physical organic chemistry. In the analytical applications of polarography only aqueous solutions or mixed aqueous/organic solvents are common. The organic component of a mixed solvent is usually added to increase the solubility of the sample or to prevent aggregation of analyte molecules. Sometimes addition of the organic component improves the shape of the wave. Organic solvents can also alter adsorption characteristics of some analytes.

SAQ 2.4a   What effect would aggregation of analyte molecules be likely to have on the polarographic behaviour of that analyte?

SAQ 2.4b   How might an organic solvent alter the adsorption behaviour of an analyte?

2.5. FUNCTIONAL GROUPS IN ORGANIC POLAROGRAPHIC ANALYSIS

Polarographic determination, by itself, is generally not specific for a particular organic compound. Usually the determination is, in practice, of a particular electroactive group and hence of a class of compounds.

For example, the half wave potentials of most benzene diazonium salts lie within about 100 mV of each other. Since at least 200 mV must separate waves before they can be sufficiently resolved, the different substituted benzene diazonium salts cannot be differentiated from one another by polarography alone. The other substituents on the benzene ring, such as methyl, hydroxyl, methoxyl, produce only relatively small shifts in half wave potential, through their electron donating or withdrawing properties. This is referred to as a substituent effect.

Substituents in a class of organic compounds can have two effects on the polarographic behaviour. Where the effect of the substituent is merely a polar electron-donating or-withdrawing effect, distant from the active group, only small shifts in half wave potential are likely. Where the substituent changes the degree of conjugation of the electroactive group, large changes of half wave potential and of the reaction mechanism itself are possible.

For example aliphatic- and aromatic-nitro compounds will have markedly different behaviour. Another example would be olefinic double bonds. Simple aliphatic alkenes are not electroactive, but increasing the conjugation by introducing two aromatic substituents on the alkene group renders it readily reduced at the dropping mercury electrode. In practice such changes in conjugation should be thought of as creating a new class of electroactive compounds.

SAQ 2.5a   If the introduction of a hydroxy substituent
             causes only a very small shift of the half wave
             potential of a particular compound, does this
             mean that a method, using polarography, cannot
             be developed to selectively determine both the
             parent compound and its hydroxy derivative?

To list all of the electroactive functional groups and classes of compound would be impossible, however the following list can act as an
introductory guide.

2.5.1. Generally Electroactive Groups or Bonds Susceptible to Reduction at the DME

carbon-carbon double bonds, if highly conjugated,

carbon-carbon triple bonds, if conjugated to an aromatic ring, some carbon-halogen bonds,

conjugated carbonyl groups,

quinones,

carbon-nitrogen double bonds,

heterocyclic compound containing two or more heterocyclic nitrogen atoms,

nitrogen-nitrogen double bonds,

nitro compounds,

nitroso compounds,

nitrosamines,

N-oxides,

S-oxides,

carbon-sulphur double bonds,

sulphur-sulphur bonds.

2.5.2. Groups or Bonds Generally not Electroactive to Reduction at the DME

carbon-carbon single bonds,

isolated carbon-carbon multiple bonds,

alkyl groups,

some aryl groups such as benzene rings,

some carbon-halogen bonds,

most alcoholic groups,

most carboxylic acid groups,

most ether linkages,

amines,

amides,

most carbon-nitrogen single bonds.

Good references sources to the wide range of electroactive compounds exist, such as that of Perrin (1965).

2.6. THE ROLE OF COMPLEXATION IN THE POLAROGRAPHIC ANALYSIS OF METALLIC IONS

At the dropping mercury electrode, each metallic ion is reduced to the metallic state at a particular characteristic potential. However this characteristic potential applies only to the uncomplexed metallic ion (strictly speaking the aquo-complex). If the metallic ion is present as a complex with some ligand, then these ligand groups must be removed prior to electrolysis. When the most stable complex forms in solution, energy is given out; this energy must be re-supplied to the complex ion to remove the ligands. This extra energy requirement is expressed in raising (to more negative voltages) both the electrode potential of the reduction and with it the half wave potential of the polarographic wave.

If the complexing agent is in sufficient excess so that the metallic ion M^{n+} is effectively totally complexed as ML_q, it can be shown that the shift in potential is given by:

$$\Delta E_{\frac{1}{2}} = \frac{-RT}{nF} \ln K_q - \frac{qRT}{nF} \ln c_L$$

In some cases the polarographic wave is due to a change of oxidation state of the metallic ion and not a reduction to the elemental metallic form, for example reduction of Fe^{3+} to Fe^{2+}. Complexation of both oxidation states in the solution must be considered. If the oxidised form makes the stronger complex with the ligand in question, the half wave potential is generally shifted to more negative potentials. On the other hand if the reduced form produces the stronger complex the potential is shifted to more positive voltages. If the product of the electrolysis forms the stronger complex this makes the electrolysis easier, while if the product forms the weaker complex this makes the electrolysis more difficult.

The above description is highly simplified and assumes a thermodynamically reversible reduction or oxidation. However it does illustrate the power of complexing agents to shift or alter the polarographic waves of metallic ions.

The importance of this to the analytical chemist is that the use of complexing agents allows adjustment of the half wave potential. If two metallic ions have similar half wave potentials each will interfere with the determination of the other. By adding a suitable complexing agent their half wave potentials can be shifted until they are far enough apart to no longer interfere. Almost any combination of metallic ions can be selectively determined by polarography by finding a suitable complexing medium.

As many sample solutions, particularly samples from environmental analysis, may contain unknown complexing agents it is probably a good idea to add a chosen strong complexing agent to ensure a known reproducible state of complexation.

Complexation can change the nature of the electrode reaction. In one medium a metallic ion might be reduced to the metal in a series of steps showing as two or more waves. With another complexing agent present, the reduction potential of the original ion and the intermediate state might be brought together to yield one single wave. Sometimes the product of the reduction in different complexing media can be quite different. For example, in a thiocyanate solution chromium(III) is reduced in a single step or wave to metallic chromium. The wave height is very reproducible and suitable for analytical application. In contrast in an ammonia/ammonium chloride buffer solution the wave is due to reduction only as far as chromium(II) ions. The great instability of the latter ion towards reoxidation in the solution adds an extra catalytic component to the wave, resulting in very poor reproducibility. The difference lies in complexation of the chromium ions in the two solutions.

Sometimes the ligand or complexing agent may itself be electroactive. In such a case the complexation with a metal ion is likely to significantly alter the polarographic behaviour of the ligand itself. In some cases the ligand may only become electroactive after it has formed a complex with a metal ion. Complexation, in such cases, can be used to render an electroinactive analyte electroactive. A good example would be certain electroinactive amino-acids which become active when complexed to copper(II) ions.

Since many of the complexing agents used are themselves weak acids or bases control of pH is most important. The pH of the solution will in such cases control the stability of the complex and in turn the half wave potential of the resultant wave. The effectiveness of a complexing agent to resolve the waves for two metallic ions will often require a particular pH range to be chosen.

∏   How does the pH control the stability of metal complexes with ligands such as weak acids?

         Usually only the anionic form of the weak acid can act as the ligand (for bases usually the neutral free basic form). The pH controls the effective concentration of the individual complexing form. It is this species concentration which controls the shift in the half wave potential.

SAQ 2.6a   If several different complexing agents are present in the solution, how will this affect the polarographic behaviour?

SAQ 2.6c

SAQ 2.6b

> If two metallic ions have similar half wave potentials as the free aquo-ion and both form complexes with a given complexing agent, would you expect the two ions to have similar half wave potentials in the presence of an equal excess of that complexing agent?

2.7. THE USE OF ELECTRODES OTHER THAN THE DROPPING MERCURY ELECTRODE

Today the term polarography is exclusively used for voltammetry at a dropping mercury electrode (or DME). The advantages of this unique electrode have already been described (Part 1). The dropping mercury electrode is the simplest and easiest to use and where ever possible it should always be chosen. However the dropping mercury electrode has two very serious drawbacks.

The dropping mercury electrode is mechanically very unstable. It requires the use of static electrolytes and the drop is very easily dislodged. The DME is quite unsuitable for on line or continuous monitoring in the presence of any moving electrolyte. Contrast this with the use of a very small graphite tube electrode which has even allowed the continuous *in situ* voltammetric monitoring of drugs and metabolic products in the blood stream of untethered living animals.

∏   What in fact holds the dropping mercury electrode on to the capillary?

Surface tension forces around the circumference of the orifice. This is obviously very weak compared to the construction of solid electrodes.

Although excellent for the study or application of electro-reduction processes, the dropping mercury electrode allows only a very narrow voltage range on the anodic or oxidation side. This is due to the relative ease of oxidation of mercury itself. For example the common analgesic paracetamol is electroinactive at the dropping mercury electrode. At a glassy carbon electrode it undergoes electro-oxidation giving a well formed wave. The use of various solid electrodes, with a large available anodic range, opens a very wide range of compounds, electroinactive at the DME, to electro-oxidation and voltammetric analysis.

The major disadvantage of solid electrodes is the great difficulty in obtaining reproducible surfaces. During electrolysis the surface is not renewed and contamination or fouling of the surface is a major problem. Problems due to adsorption phenomena are much greater at a solid electrodes than at a dropping mercury electrode. Solid electrodes should be avoided unless the dropping mercury electrode cannot be used.

The most commonly used solid electrodes in the past were platinum and gold, but in recent years various graphite based electrodes have become dominant. Common types of graphite electrodes include, graphite paste (very fine graphite powder suspended in a grease or wax), glassy or vitreous carbon, and solid graphite impregnated with wax. In all cases the graphite must be rendered nonporous. Even graphite powder suspended in resins such as Araldite has its applications. In all cases the solid electrodes must be finely polished before use and receive regular polishing during use.

Π What general chemical characteristic do gold, platinum and graphite have in common?

They are chemically very inert and do not readily react with solution components. Mercury is also very inert at cathodic (reduction) potentials. It is however much less inert towards oxidation and this is why it has such a small working range on the anodic or oxidation side.

The main advantages of graphite are its wide working potential range, the relative ease with which it can be polished etc and its chemical inertia. Graphite, platinum and gold all have a similar anodic potential range typically up to $+1.5$ V in aqueous media or up to $+2$ V in organic solvents. With mercury the anodic limit is usually significantly less than $+0.5$ V. The cathodic range for graphite up to -1 V compares quite well to mercury but for platinum and gold the cathodic limits are about -0.4 V and -0.7 V respectively.

The major difference between polarography at the dropping mercury electrode and voltammetry at solid electrodes is that every point in the current potential curve for the DME is measured for a fresh electrode in an almost undepleted solution. With the solid electrode, as each potential in the voltage scan is reached the effects of depletion at previous potentials are still felt.

SAQ 2.7a What is the major effect of depletion of the analyte on the voltammetric signal?

SAQ 2.7a

If the voltage scan rate is fast then we speak of fast linear sweep voltammetry. This technique gives rise to peaks rather than waves. As the potential rises at first no current will flow until the minimum deposition potential is reached. The current will then rise rapidly but the resultant electrolysis causes major depletion of the analyte in the electrode vicinity. Thus the current will decrease giving the peak shown in Fig. 2.7a.

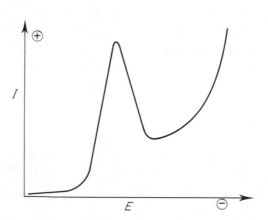

Fig. 2.7a.   *A typical fast linear sweep voltammetric peak*

Fast linear sweep voltammetry can give a complete current potential curve in times of a second or much shorter and is thus suitable for following the kinetics of chemical reactions or where ever very rapid analysis is necessary. For more conventional polarographic methods scan times of 10 to 20 minutes are more typical. However the fast scan rates of fast linear sweep voltammetry result in large charging or capacitive currents necessary to charge up the electrode surface to the potential required. This results in a loss of sensitivity etc,in consequence fast linear sweep voltammetry is not a popular analytical technique.

SAQ 2.7b	Why should a fast scan rate produce a large charging or capacitive current at a solid electrode?

The chief interest in fast linear sweep voltammetry probably lies in the significant build up of electrolysis product on or at the electrode. If the potential scan is reversed before the electrolysis product has escaped, it is possible to examine its electrochemical behaviour. If the original electrochemical reaction is thermodynamically reversible the reoxidation (or re-reduction) peak should occur almost at the same potential (separated by $59/n$ mV). For a thermodynamically irreversible reaction the reoxidation (or re-reduction) peak of the product is either absent or widely separated from the original (Fig. 2.7b).

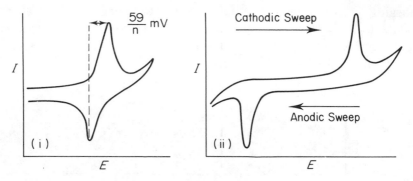

Fig. 2.7b.   *Cyclic voltammogram showing peaks for (i) a reversible, and (ii) an irreversible process*

This fast linear sweep voltammetry with periodic reversal of scan direction is better known as Cyclic Voltammetry. Cyclic voltammetry is a vital tool to the physical electrochemist but has little or no direct electroanalytical application.

SAQ 2.7c   How also might the reversibility of an electrode process be investigated?

SAQ 2.7c

Voltammetric analysis at solid electrodes usually involves slower scan rates, typically 5–10 minutes per volt. In addition the solution is often stirred. Both of these factors lessen the depletion effect and the peak in the current potential curve is much less pronounced and the signal is more wave like (Fig. 2.7c). The precise shape depends on the scan and stirring rates. With a fast enough stirring rate the conventional dc polarographic wave is attained.

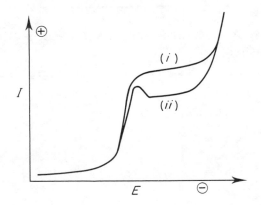

Fig. 2.7c. *Voltammetric current/potential curves at a solid electrode for an analyte using (i) a faster and (ii) a slower stirring rate*

The limiting current is given by:

$$I_{lim} = K'nFADc$$

where A is the effective surface area of the electrode, and K' a constant, the mass transfer coefficient.

Thus the limiting current, as the wave height, is proportional to the concentration of the analyte, and can be used for analyical application.

The constant K' is a mass transport coefficient, which depends on cell geometry, the electrolyte and above all on the rate of stirring. The stirring of the solution must be precisely controlled and reproducible. Often the stirring is carried out by rotating the electrode.

The design of solid electrodes varies enormously. Small wires or spheres may be employed. A 'solid' mercury electrode could be made by picking up a mercury film on such wires or spheres. Rotating disc electrodes are very popular. These consist of a rod of metal or graphite in a insulated sleeve. The rod is then cut to expose a flat disc which can be easily rotated with the rod as axis. A new surface can be easily obtained by grinding followed by polishing or by cutting and polishing.

| SAQ 2.7d | Why must the electrode surface be regularly polished? |

More complicated electrodes have been designed to act as continuous flow sensors. A tubular electrode is very suitable for flow injection analysis. In this, the sample is injected into a stream of buffered electrolyte flowing in a narrow tube system, through the tubular working electrode allowing a voltammetric determination. Reference and auxilary electrodes are down stream of the working electrode. Such a system requires a very small sample volume, is very sensitive and allows a rapid rate of analyses. The so called wall jet cell (Fig. 2.7d) has a similar use and consists of a thin layer cell in which the solution enters through a fine nozzle and impinges on the centre of a glassy carbon disc electrode. This offers a very small cell volume (0.5–5 μl dead volume) and a very high mass transport of analyte to the electrode surface ensuring a high signal and sensitivity. Such refinements would be impossible with the dropping mercury electrode.

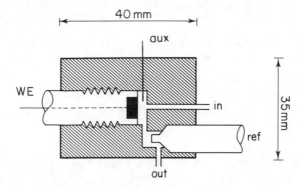

Fig. 2.7d. *A wall jet cell. Working electrode (WE) is glassy carbon; reference electrode (ref) a saturated calomel electrode and the auxilary electrode (aux) a platinum wire. The cell body is made of plexiglass*

SUMMARY AND OBJECTIVES

Summary

A diffusion-controlled limiting current is usually required to obtain a precise determination of the analyte concentration in a solution.

The presence of other kinds of limiting current and adsorption effects pose a hazard to precise and accurate polarographic analysis.

The choice of supporting electrolyte allows conditions to be chosen where the wave of analytical interest can be diffusion-controlled with a wave height independent of pH etc. Choice of solvent, control of pH and state of complexation are often used to separate the waves due to interfering species.

The many advantages of the dropping mercury electrode make it the most important and widely used electrode.

Objectives

You should now be able to:

● understand the terms reversible and irreversible and their significance in the context of voltammetric analysis;

● recognise the importance of a diffusion-controlled limiting current in obtaining a precise determination of the analyte concentration in the solution;

● recognise other kinds of limiting current on the basis of experimental data and know why they are unsuitable for analytical applications;

● identify the presence of adsorption effects and recognise the hazard they pose to a precise and accurate polarographic analysis;

● explain how the supporting electrolyte, for a particular analytical purpose, is chosen with respect to solvent, complexing agents and pH etc;

● recognise the importance and advantages of the dropping mercury electrode, but also be aware of other electrode types and the advantages of the techniques that can be employed with them.

3. Pulse Polarographic Techniques

Overview

In this part the factors limiting the sensitivity of classical dc polarography are discussed. Two more modern techniques, normal and differential pulse polarography, are introduced, these use a more complex applied voltage signal and current sampling technique to minimize interfering background signals and to maximize the analytical signal, thus giving a greatly increased sensitivity. The principles involved are discussed in a non-mathematical manner.

3.1. INTRODUCTION

Pulse polarographic techniques were developed primarily in the general trend towards greater sensitivity in analytical methods as ever lower limits of detection and determination were demanded by the user disciplines. For example, dc polarography was quite suitable in sensitivity for the analysis of electroactive drugs in pharmaceutical preparations, but far too insensitive to determine the same drugs in biological fluids during pharmacological investigations. This was particularly frustrating because polarography showed the ability in many cases to differentiate the drug from its metabolites. The pulse polarographic techniques were in fact developed in the late 1950's but their widespread application was for a long time severely hindered by the cost and difficulties of the instrumentation. However, with the revolution in the electronic sciences in the last decade, reliable pulse polarographic equipment has become commercially available offering a sensitivity at least as good as its main rivals (atomic spectroscopy aas and hplc) at a much lower cost. Today, pulse techniques are the most widely used methods in analytical polarography. However they do not just offer increased sensitivity but have other advantages to offer as will become apparent.

3.2.   THE LIMITATIONS OF DC POLAROGRAPHY

The sensitivity of most instrumental techniques is limited by the signal to noise ratio. (See also 1.5.2).

∏   In general in analytical instrumentation (i) what is meant by noise and (ii) what is the usually accepted minimum signal to noise ratio?

   (i)   Unwanted instrumental signal unrelated to the analyte or its concentration, often random or uncertain.

   (ii) A minimum signal to noise ratio of 3 : 1 is generally considered necessary to differentiate signal from noise with certainty.

∏   What is meant by Faradaic current?

      The electrolysis cell acts like a electronic component in the circuit with its own electrical resistance, capacitance etc. When a voltage is applied a certain current will flow. This current, and the conductivity it represents, is made up of several components. The most useful is that caused by simultaneous electron exchange (oxidation and reduction) involved in the electrochemical reaction at the two electrodes. Since this latter current obeys Faraday's Law of Electrolysis it is often refered to as the Faradaic current.

With classical dc polarography the signal is the mean Faradaic current and the principal source of noise is the capacitive current for the growing mercury drop.

For diffusion-controlled processes the faradaic or electrolysis current grows during the drop lifetime $I = kt^{\frac{1}{6}}$ (Fig. 1.5d). This growth is the resultant of two opposing processes. The first is the increase in the size of the growing mercury drop causing an increase in the current. The second is the depletion of the electroactive species from the solution and growth of a depletion zone around the electrode, due to electrolysis. The latter produces a decrease in the current; Fig. 3.2a shows this decrease in current for an electrode of fixed size.

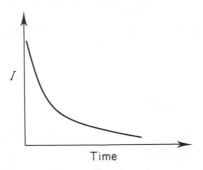

Fig. 3.2a. *The decrease of diffusion current due to depletion at a stationary electrode*

∏ What is the relationship between electrical charge and potential?

The electrode/solution interface acts like a capacitor. The higher the charge density on the surface, the higher the potential. If the surface area increases more charge is required to maintain the density. That extra charge flowing represents a current.

The capacitive current (Fig. 3.2b) falls during the lifetime of a drop. The capacitive current required to charge a solid electrode to the desired potential would decay much more rapidly than is the case for the growing DME. However the mercury drop electrode is increasing in surface area as it grows and so an extra charging current is required just to maintain the potential. This current decreases with time since the rate of the growth of the mercury drop falls rapidly, almost ceasing before the drop falls off.

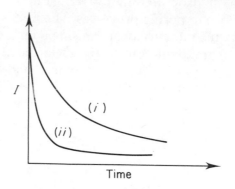

Fig. 3.2b.   *Comparison of the capacitive current on (i) the DME, and (ii) a stationary electrode*

The signal to noise ratio, that is the ratio of the Faradaic current (proportional to concentration) to the capacitive current is greatest at the end of the drop lifetime. Classical dc polarography records the mean current during the drop lifetime and thus does not record the maximum Faradaic current. Worse still, it includes an appreciable contribution from the capacitive current. This effectively limits the sensitivity of the classical dc technique to a limit of detection of about 5×10^{-5} mol dm^{-3}.

A somewhat better sensitivity could be obtained if the current were recorded at the end of the drop lifetime, thus maximising the signal to noise ratio. This is the basis of sampled dc polarography in which the current is measured during the last 15% of the drop lifetime. The recorder is kept stationary during the first 85% of the drop lifetime, known as the blanking time. However this modification represents only a relatively small increase in sensitivity (typically about 1×10^{-5} mol dm^{-3}) over the classical method. The chief loss in sensitivity remains in the depletion of the electroactive species in the vicinity of the electrode (in a developing depletion zone) during the first 85% of the drop lifetime. The concentration of the analyte in this zone may typically be reduced by an order of magnitude before the current is measured (Fig. 3.2a).

SAQ 3.2a Sum up, in two key words, the factors limiting the sensitivity of dc polarography.

3.3. NORMAL PULSE POLAROGRAPHY (NPP)

A great improvement in sensitivity would be possible if electrolysis could be prevented during most of the drop lifetime, that is the potential kept at a lower voltage until the moment of measurement. This is the basis of normal pulse polarography. The potential is kept at a suitable constant base potential throughout the drop lifetime. The chosen potential signal is imposed as a very short pulse (about 60 ms) near the end of the drop lifetime (Fig. 3.3a). Thus very little electrolysis and depletion occurs. Typically the Faradaic current from normal pulse polarography is about ten times that from classical dc polarography.

The problem still remains of the capacitive current. When the voltage pulse is imposed a significant capacitive current will be required to produce this potential. However near the end of the drop lifetime the growth of the surface area of the drop has almost ceased, particularly relative to the short length of the pulse, and the electrode is almost stationary. As a glance at Fig. 3.2b will show, this results in the capacitive current decaying rapidly, much more rapidly than with the classical technique. The current measurement is therefore taken in an even shorter pulse (about 15 ms) near the end of the potential pulse once the capacitive current has decayed to a low steady value (Fig. 3.3a).

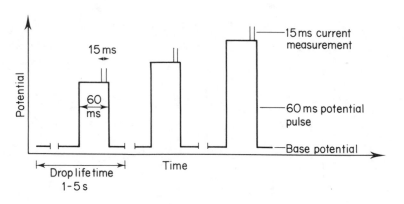

Fig. 3.3a.   *The profile of the potential pulse and current measurement in normal pulse polarography*

The overall form of the applied voltage signal is of a series of potential pulses, one to each drop, rising in a linear ramp with a base potential maintained between the pulses. The current recorded at the end of each pulse is recorded and plotted against the potential of the pulse.

The resultant current/potential curve is similar in form to the classical dc polarographic curve with equivalent half wave potential and limiting current (Fig. 3.3b). In practice the curve is made up of very short flat segments and is of a clearer, 'cleaner', less noisy form. The height of the pulse polarographic wave, the analytical signal, is about ten times greater than that of the classical wave height, suggesting only a similar increase in sensitivity. However, the noise, ie the capacitive current, has also been drastically reduced. Thus the sensitivity of the technique has been improved by about two orders of magnitude with a limit of detection of about 10^{-7} mol dm^{-3}.

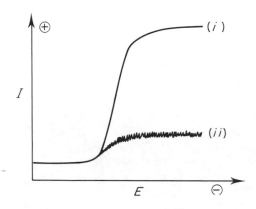

Fig. 3.3b. *Comparison of (i) the normal pulse and (ii) classical dc polarograms from the same solution at the same recorder and sensitivity*

Thus by using more sophisticated electronics and the imposition of a more complicated potential signal a much simpler current/potential curve can be obtained with the interferences of depletion and the capacitive current effectively minimised.

∏   In normal pulse polarography has the capacitive current been eliminated from the final signal?

It is not eliminated, just greatly reduced. The drop continues to grow very slowly and even with a stationary electrode capacitive current never totally reaches zero.

SAQ 3.3a   For a drop lifetime of 2 seconds, what fraction
                  of the drop lifetime is taken by the pulse?

SAQ 3.3b   Why does the normal pulse polarographic wave
                  in Fig. 3.3b not display the rapid current flucta-
                  tions found with the classical dc wave?

SAQ 3.3c   If the potential pulse is to occur at the exact same point in the drop lifetime, how might the pulse be synchronised to the dropping of the mercury electrode?

3.4.   DIFFERENTIAL PULSE POLAROGRAPHY (DPP)

This is the most widely used polarographic technique today. It differs from normal pulse polarography in that after the potential pulse the potential does not return to a constant base value. Instead the potential pulse itself is of a small constant amplitude (10–100 mV) and is superimposed on a conventional rising linear dc voltage ramp. Once again the pulse is imposed for about 60 ms near the end of the drop lifetime when the growth of the drop has almost ceased (Fig. 3.4a).

The current is measured in two intervals of about 15 ms, the first immediately prior to the potential pulse and the second during but towards the end of the potential pulse. The final current signal displayed is in fact the difference of these two current values.

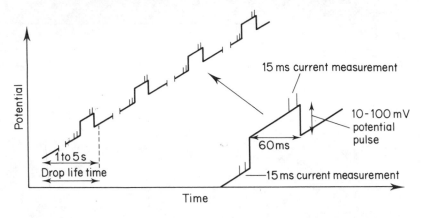

Fig. 3.4a.   *The profile of the potential pulse and current measurement in differential pulse polarography*

The two current values represent the current at two potential values separated by about 10–100 mV (the pulse amplitude). This difference in current will be greatest on the steep rising part of the polarographic wave around the half wave potential, where a small change in potential produces a large change in current. Thus this technique in fact produces not a wave but a peak with the highest current signal at roughly the half wave potential of the classical dc and normal pulse polarography (Fig. 3.4b). Since the output signal increases with the steepness or slope of the conventional current potential curve, this final curve approximates to a derivative or differential of the classical polarographic current potential curve.

∏   Why has the expression 'approximates to a derivative' been used in the above paragraph?

Some dc polarographs have a so called derivative circuit. This, in fact, attempts to give, purely by electronic means, a true mathematical derivative of the classical dc polarographic curve produced by the electrolysis cell. Differential pulse polarography does not give this because electrolysis conditions have been altered in the cell itself. For example major depletion, throughout the drop lifetime, has been occurring for the potential before the pulse but the extra current during the pulse has had no time to produce serious depletion. Thus differential pulse polarography does not give a true derivative of the classical wave form. The dc derivative circuit offers little advantage since it is limited to the original classical dc polarographic data from the cell, and thus is of poor sensitivity. It merely juggles mathematically with the dc data. Differential pulse polarography brings improved data from the cell itself.

The potential of the peak E_p is indicative of which species is involved. If the reduction (or oxidation) mechanism is diffusion-controlled the concentration of the species controls the Faradaic current. Since differential pulse polarography effectively displays the derivative of this current, theoretically it is the area under the peak which is proportional to the concentration. However, provided the shape of the peak does not change, the height of the peak is also proportional to concentration. The choice between the two modes of measurement will be discussed later.

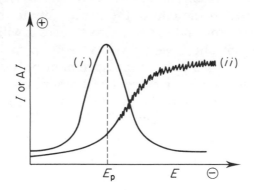

Fig. 3.4b. *Comparison of (i) the differential pulse, and (ii) classical dc polarograms from the same solution at the same recorder sensitivity*

Note that the rising linear voltage ramp in differential pulse polarography is exactly similar in form to that of classical dc polarography, except for the superimposed pulses. This means that electrolysis and hence depletion occurs throughout the drop lifetime. Thus in differential pulse polarography the current signal is greatly reduced compared to that in normal pulse polarography and is more similar to that in classical dc polarography. At first glance this would, wrongly, suggest that the differential pulse technique should be inherently less sensitive than normal pulse polarography. In fact differential pulse polarography is typically an order of magnitude more sensitive than the normal pulse mode. Typical limits of detection are 10^{-7}–10^{-8} mol dm^{-3} for the differential pulse technique and 10^{-6}–10^{-7} mol dm^{-3} for the normal pulse technique while for the classical dc polarography it would be only about 0.5×10^{-4} mol dm^{-3}.

How can there be an increases in sensitivity if the signal has been reduced by depletion? Clearly differential pulse polarography must reduce the noise, in the signal to noise ratio. It must show better resolution. At low concentration levels the favourable signal to noise ratio of differential pulse polarography gives well defined peaks where no dc response can be obtained.

Although depletion occurs throughout the drop lifetime, the current measurement is only taken in very brief pulses at the end of the drop lifetime. At the end of the drop lifetime the drop has almost stopped growing and the capacitive current will be reduced to its almost constant and lowest level. The jump of 10–100 mV occuring between the two current measurement pulses will have very little effect on the capacitive current and other non-faradaic sources of noise. On the other hand, the small potential jump will produce a large change in the Faradaic current particularly at the peak potential. It is the change in current, on either side of the potential pulse, which differential pulse polarography records. Thus the differential pulse mode allows the maximum differentiation of the Faradaic or analytical signal from the background signal.

Resolution is in fact a keyword for differential pulse polarography. In classical dc polarography at least 200 mV or more are required between half wave potentials before interfering or overlapping waves can be resolved. With differential pulse polarography overlapping peaks can be usefully resolved for analysis if separated by as little as 50–100 mV. These values are only relative and are larger for broader and less well formed peaks or waves.

SAQ 3.4a Summarise in one sentence each, for both methods, the potential signal imposed and the current signal measured for normal and differential pulse polarography.

SAQ 3.4b Is the sensitivity of the recording apparatus likely to be of importance to the overall sensitivity of the polarographic method?

3.5. CHOICE OF OPERATING PARAMETERS IN PULSE POLAROGRAPHY

With the imposition of a more complicated voltage signal in the pulse techniques, some new operating parameters must be set on the instrument in addition to those of classical dc polarography. These are the base potential in normal pulse polarography and the pulse amplitude in differential pulse polarography.

3.5.1. Choice of the Base Potential in Normal Pulse Polarography

Normal pulse polarography is somewhat unwisely neglected today and there is a general tendency to go from classical dc polarography straight to differential pulse polarography. The latter is a little more sensitive than the normal pulse technique (by about one order of magnitude). However the normal pulse mode has some very distinct advantages, and these lie in the choice of the base potential to which the electrode returns between pulses.

Probably the biggest problem in analytical polarography is adsorption of species on to the surface of the electrode. This can be adsorption of the analyte, its electrolysis product, or any other species from the solution. The effects of adsorbed species can be very varied indeed. They can produce the splitting of polarographic waves, the distortion of their shapes, shifting of the half wave potentials, depression or even elimination of the wave heights, etc. The adsorbed forms may produce small waves of their own, known as pre-waves or post-waves, separate from the main diffusion controlled wave. On the other hand some adsorbed species have little or no effect.

If adsorption of the electrode product is likely to cause interference then this can be minimised in normal pulse polarography by setting the base potential to a low value at which no product is formed. A very small amount of electrode product will be formed only during the short duration pulse. This will cause a much smaller interference than would occur in classical dc polarography or in differential pulse polarography, since with the latter two techniques electrode product is formed throughout the drop lifetime.

As the potential of the electrode changes so the stability of adsorbed forms varies. An individual species can have several adsorbed states each of which occurs only over a particular electrode potential range. As the potential moves out of this range, the species is either desorbed completely from the surface, or reorientates into a new adsorbed state.

∏ How can one chemical species have several adsorbed states?

Larger more complicated molecules can bind to the surface of the electrode at different angles and different orientations with different active groups involved in the adsorption process. Each represents a different adsorbed state. Each is usually stable at a different range of electrode voltages since each is of different energy.

In normal pulse polarography the base potential can be set to a value at which the interfering species is either desorbed (ejected) from the electrode surface for the bulk of the drop lifetime, or is in the least interfering adsorbed state. This option is not open to the other common polarographic modes.

∏ What might be done to minimise interference from adsorption effects in these other polarographic modes?

The adsorption properties of a given species can change greatly from one solvent or electrolyte to another. Changes in the protonation of the species can also alter the properties. The supporting electrolyte might be changed until one is found in which the sample shows the least adsorption interferences. Some surface active agents have the property of offering little interference themselves with the wave or peak but are able to oust other interfering adsorbed material from the surface. Triton X-100, a polyalcohol, is often added in very small quantities (0.001%) to this end.

In some normal pulse polarographic equipment the base potential to which the voltage returns between the pulses is simply equal to the initial potential of the potential scan. The initial potential of the scan is the potential imposed during the first pulse. The potential during subsequent pulses increases at a regular interval from this initial pulse to form the linear potential scan. In some instruments the initial and base potentials can be set separately. Note it is possible to scan towards the normal pulse polarographic wave from a base or initial potential either above or below the wave.

Thus it can be seen that normal pulse polarography is potentially the least sensitive to interference from adsorption effects of the common modes of polarography.

3.5.2.   Choice of Parameters in Differential Pulse Polarography

Before analysis a choice has to be made of the potential pulse amplitude imposed. With some differential pulse instruments a fixed pulse amplitude of, usually 50 mV, is the only possibility. Others allow a continuous choice 10 mV to 100 mV.

The peak current I_p for a totally thermodynamically reversible electrode process controlled by diffusion has been derived by Parry and Osteryoung.

$$I_p = \frac{n^2 F^2 A c}{4RT} \left(\frac{D}{\pi t} \right)^{\frac{1}{2}} \Delta E$$

where ΔE is the amplitude of the potential pulse.

SAQ 3.5a	What is the equivalent equation for classical dc polarography?

SAQ 3.5a

SAQ 3.5b What is meant by thermodynamically reversible or irreversible?

The Parry-Osteryoung equation shows immediately that the height of the peak is proportional to the concentration as is necessary for analytical use. It also shows that the peak height is proportional to the potential pulse amplitude. That is the higher the amplitude the greater the sensitivity. However increases in pulse amplitude result in broadening of the peaks and subsequent loss of resolution. Two close lying peaks will not be resolved unless the pulse amplitude is significantly smaller than the separation in the two peak potentials.

Thus the choice of potential pulse amplitude must be a compromise between a higher value for increased sensitivity and a lower amplitude for increased resolution. This is particularly true for thermodynamically irreversible processes which produce broader, lower and less well formed peaks than do reversible processes.

The scan rate (mV s^{-1}) must be chosen carefully. The differential pulse peak consists of a series of straight line steps as the pen is held stationary (in the current direction) between the pulses. If the scan rate is too high the steps will be too coarse for adequate resolution. The slowest scan rate gives the best results but a reasonable compromise with analysis time may be necessary.

A choice has to be made for calibration and measurement between use of peak height or of the area under the peak. Since differential pulse polarography is a differential or derivative technique it is the area under the peak which is proportional to the current of electrolysis and hence to concentration. The measurement of a peak height is much more convenient. However the relationship between peak height and area is only constant as long as the shape of the peak does not alter. This is probably the case in most analytical problems but it is not always so. Since irreversible processes produce lower broader peaks than reversible processes, any change in thermodynamic reversibility will alter the relative peak height. The greatest problem is adsorption onto the electrode surface.

Differential pulse polarography is particulary susceptible to surface active phenomena. Adsorbed forms of the analyte and its electrode products can give rise to separate peaks. But even the adsorption of otherwise inactive third species can alter the reversibility and electrode kinetics of the process producing sometimes huge changes in the shape of peak. The area under the peak will, however, remain constant in most cases. But the height of the peak is of no use. If surfactants are likely to be present it is best to calibrate and measure the area under the peak.

On the other hand when two peaks partially overlap (Fig. 3.5a) it is very difficult to separate the area under one from that under the other. The two peak heights may be well defined, and provided the tail of the other peak would effectively be zero at this potential, then the peak heights can be used with much greater precision than the areas under the peaks.

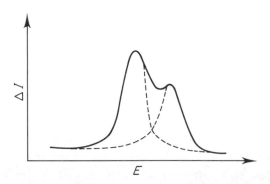

Fig. 3.5a. *The problem of differentiating the area under two overlapping peaks in differential pulse polarography*

The area under the peak should be used to minimise effects such as adsorption on the peak shape, while the peak height should be used to minimise interference from other overlapping peaks. There is no correct method. Both types of calibration should be tried with known typical samples and the more reliable chosen.

SAQ 3.5c | What methods could be used to actually measure the area under the differential pulse peak on the chart recorder output?

SUMMARY AND OBJECTIVES

Summary

The chief factors limiting the sensitivity of dc polarography are noise due to the capacitive current (charging the mercury drop) and reduction of the analytical signal through depletion of analyte in the vicinity of the electrode by electrolysis throughout the drop lifetime. Normal pulse polarography achieves an increased sensitivity by confining the electrolysis to a short pulse at the end of the drop lifetime and by minimizing the capacitive current by sampling the current signal in an even shorter pulse once the capacitive current has decayed to a minimum.

Differential pulse polarography gains greater sensitivity by taking advantage of the fact that a small jump in potential has a much greater effect on the analytical signal than on sources of noise such as capacitive current. A differential technique based on a small potential pulse thus maximizes the signal/noise ratio leading to greater sensitivity.

Objectives

You should now be able to:

- understand the factors limiting the sensitivity of dc polarography;

- understand how normal pulse polarography can achieve a so greatly increased sensitivity;

- explain the even greater sensitivity of differential pulse polarography;

- explain how the operating parameters are chosen for pulse polarographic measurements. (That is you should understand in general terms what each knob on a typical pulse polarograph is for.)

4. Stripping Voltammetry

Overview

The sensitivity of any analytical technique can be greatly increased by introducing a preliminary pre-concentration step, eg solvent extraction. In stripping voltammetry an electrochemical preconcentration technique is used. The analyte is concentrated, from very dilute solutions, by electrolysis to an insoluble product which collects at the electrode and can be subsequently determined with a very high sensitivity. The method is applicable only to a limited number of important analytes. Stripping voltammetry requires the use of solid or stationary electrodes, (2.7).

4.1. INTRODUCTION

Although differential pulse polarography is the most sensitive direct polarographic technique, an even greater sensitivity can be obtained by employing stripping voltammetry. This latter technique involves a preconcentration step before the final voltammetric determination. It is the preconcentration step which allows the great sensitivity of stripping voltammetry.

The preconcentration or deposition step consists of the controlled electrodeposition, at a fixed potential, of the species of interest onto a stationary electrode. This is followed by the determination step which consists of electrolytically stripping the deposited species back into the solution.

A typical example would be in the determination of copper ions. The potential of the electrode is set sufficiently negative to reduce the copper ions to metallic copper which is deposited onto the electrode. The electrolysis is allowed to continue for an appreciable time collecting and concentrating copper from a relatively large volume onto the small electrode. Then, in the next stage, the potential is moved towards more positive (anodic) values and the copper is stripped off the electrode and reoxidised to copper ions giving the analytical signal. The long preliminary deposition of copper on the electrode makes a much higher quantity of copper available, to the determination step, than would be available from the low quantities of copper ion in the original solution, in the vicinity of the electrode.

As the potential is taken to more positive potentials, in the stripping step, initially no current will flow since the potential is still too negative to allow dissolution of the copper metal. When the dissolution potential is reached the copper metal will begin to dissolve and the current will rise exponentially. However there is only a finite limited quantity of copper metal available and as it is used up so the current must fall back to zero or in practice the base line. Thus the stripping signal takes the form of a peak as shown in Fig. 4.1a.

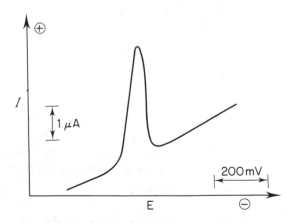

Fig 4.1a.   *A typical linear sweep stripping voltammetry peak*

∏   What could cause the rising slope of the baseline?

This is a residual current largely due to capacitive current.

The key to the sensitivity lies in concentration, at the electrode, of the product of the deposition step. Clearly this product must be insoluble in the solution otherwise it would simply diffuse away back into the bulk of the solution and would not concentrate or build up at the electrode surface. The product may collect as a deposit or adsorb at the electrode surface or if a mercury drop electrode is used the product may dissolve in the drop as an amalgam.

∏   Which type of electrode product will dissolve to form an amalgam?

Only metals will do this.

Obviously a dropping mercury electrode (DME) cannot be used for stripping voltammetry or the falling drop would carry away the product. Generally a stationary mercury drop or a solid electrode is used.

Since the procedure involves both a reduction and an oxidation step the electrode reactions must be capable of being reversed, although they need not be strictly thermodynamically reversible.

The above gives a basic description of the technique but there are many variations possible on both the deposition step and the determination or stripping step.

4.2. THE DEPOSITION STEP

4.2.1. The Choice of Electrode

The variations possible are concerned with the nature of the electrode and the stripping of the solution. The most common technique is to use a stationary electrode in a stirred solution.

The most popular electrodes are the hanging mercury drop electrode (HMDE), the mercury film electrode (MFE) supported on gold or platinum, and graphite electrodes including graphite paste or glassy carbon. The mercury electrodes offer the advantage, in metal ion analysis, that the deposition product, the metal, dissolves in the mercury to form a liquid amalgam. This offers better reproducibility than a solid metal deposit on a solid electrode. In general terms the mercury film electrode is capable of more sensitive measurements than the hanging mercury drop electrode, but is less suited to relatively higher trace amounts, since the solubility capacity of a film is less. A general guide would be to use the hanging mercury drop for metal ion levels above 1 ppb and the mercury film below this.

∏   How might a mercury film electrode be prepared?

There are two ways. A very well degreased and platinum or gold electrode will pick up a mercury film on its surface. It is also possible to electrodeposit a mercury film on these platinum, gold or graphite electrodes from a solution of mercury ions.

∏   Can you think of an advantage of using a graphite electrode?

Graphite electrodes are mechanically more stable than the HMDE. A mercury electrode could not be used for the stripping voltammetric determination of mercury ions, while graphite could be used. Graphite has a wider cathodic (reduction) working range than gold or platinum and a much wider anodic (oxidation) working range than mercury (see 2.7). Graphite is an excellent substrate to hold a mercury film for MFE. Many types of high quality graphite are now available.

4.2.2. To Stir or Not to Stir

Stirring the solution during the deposition step increases the rate at which the analyte reaches the electrode to be deposited. Control of the stirring is vital. The stirring must be uniform and at a rigidly controlled rate. The position of the electrode within the cell and hence in the solution flow pattern must be absolutely reproducible. The stirring must be gentle otherwise unpredictable eddy effects will occur.

Deposition in a still, unstirred solution might appear to offer a much higher reproducibility of conditions but at a cost of a much reduced sensitivity or very much longer deposition times. Unstirred solutions are rarely used except in combination with a differential pulse stripping technique whose greater sensitivity to some extent offsets this reduced sensitivity.

∏   In an unstirred solution how would the analyte reach the electrode for deposition?

By diffusion. This is much slower than transport in a stirred solution but is also very reproducible.

An alternative to stirring the solution is to rotate the electrode. For this the electrode usually consists of a rod cut to expose a flat disc. The rod is then rotated about its axis. Such an electrode is usually made of glassy graphite, platinum or gold. sometimes a film of mercury is deposited on the electrode by electrodeposition from a separate solution of mercury(II) ions.

More unusual ideas involve the flow of solution through tubular electrodes or the use of so called 'wall jet' electrodes in which a rapid jet of solution is impinged directly onto a small electrode. Since these methods involve flow of solution through tubing rather than an open cell, they introduce the possibility of automated analysis. The samples could be automatically injected into a continuous flow of fresh supporting electrolyte – flow injection analysis. Conventional polarography does not lend itself as easily to automation as does this version of stripping voltammetry.

SAQ 4.2a	Why would the HMDE and the DME be unsuitable for use in such flow systems?

SAQ 4.2a

4.2.3. The Choice of Deposition Time

A further decision is how long to continue the deposition step. The greatest sensitivity would clearly be obtained by carrying on the deposition process until all of the analyte would be deposited on the electrode. Indeed this is sometimes done using a small volume of sample solution. However it is more usual to use a large volume of solution and only deposit a small fraction of the analyte.

It is best to avoid long deposition times. This often leads to various complications resulting in a loss of proportionality between final signal and the concentration of the analyte. One problem can be reactions of the deposit or changes in its nature over a relatively prolonged time. Too much deposit on the electrode can also cause problems.

In general a good guide is too choose a deposition time so that only about 2% of the total analyte is depleted from the solution. This allows the nature of the solution to remain essentially unaltered. This avoids changes in the deposition process that occasionally occur when depletion approaches completion.

SAQ 4.2b	A 20.0 cm^3 sample of effluent contains approximately 10^{-8} mol dm^{-3} of copper ions. At the chosen electrode this generates about 1 nA of current during the deposition step. What length of deposition time would be required to deposit about 2% of the copper?

4.2.4. The Choice of Deposition Potential

The final decision is the choice of the constant potential used in the pre-electrolysis or deposition step. Usually a potential is chosen a few hundred millivolts larger than the polarographic half wave potential of the analyte.

∏ Why would a potential equal to the polarographic half wave potential not be used for the deposition step?

At the half wave potential only half of the limiting current is flowing. The maximum limiting current, on the upper plateau of the wave, is about 50–100 mV higher. It is safer to go about 200 mV or more about the half wave potential.

The potential chosen allows a degree of selectivity. In the analysis of a solution containing a number of metal ions, each metal ion will have its own individual deposition potential. Thus only one metal or a group of metals can be deposited, avoiding the deposition of other metals which might interfere with the stripping step. The higher the potential, the more types of metal etc will be deposited and the more interferences likely.

Occasionally it might be possible to differentiate different oxidation states by choosing a potential at which only one form deposits to the metal. This will only be possible if the higher oxidation state is reduced at a more negative potential than the lower. An example would be arsenic species; arsenic(III) is reduced to elemental arsenic at much lower potentials than is arsenic(V). Dependent on substituents arsenic(V) is either electroinactive or is reduced at much more negative potentials. However with most metals it is the most negative wave which has the metal as its final product and stripping voltammetry will not differentiate the higher and lower oxidation states.

4.3.   THE STRIPPING STEP

A number of different stripping procedures have been divised but only two are of great significance – dc or linear sweep stripping voltammetry, and differential pulse stripping voltammetry. These require exactly the same instrumentation as dc polarography and differential pulse polarography.

4.3.1.   Dc or Linear Sweep Stripping Voltammetry

Dc or linear sweep stripping voltammetry is the simplest form of stripping voltammetry. This involves the imposition of a simple linear voltage scan on the electrode.

∏   What is meant by a linear voltage scan?

The voltage increase (or decreases) at a constant unvarying rate with respect to time. A plot of potential (volts) against time would be a straight line.

SAQ 4.3a | What is usually meant by the terms anodic and cathodic and when is an electrode an anode or a cathode?

(a) dc anodic stripping voltammetry. For metal ions a cathodic deposition potential (negative) is set for the deposition time reducing the ions to metal. Then the linear voltage scan is started with the potential moving towards anodic potentials (positive) for the reoxidation of the metal. A typical scan rate would be 50–200 mV s^{-1}. Since the stripping step is an oxidation or anodic step this is known as dc anodic stripping voltammetry.

∏   How does the above scan rate compare to that used in dc polarography?

It is very much faster. In dc polarography a scan rate of 2 mV s^{-1} or 100 mV min^{-1} would be typical.

(*b*) dc cathodic stripping voltammetry. For anionic species such as sulphide the preelectrolysis step (oxidation) is anodic and the stripping step is a voltage scan to cathodic potentials (negative). Hence this is dc cathodic stripping voltammetry.

As the potential scan begins no current initially flows. When the re-oxidation potential (or re-reduction potential respectively) is reached the current rapidly rises. However there was only a fixed amount of material deposited in the deposition step. The current must therefore fall back towards the base line as the last of the deposited material is reoxidised (or reduced). The dc stripping signal thus consists of a peak. Fig. 4.3a shows both the imposed voltage form and the current signal measured.

The height of the peak is used to determine the concentration in the original solution. However the peak height is dependent on both the concentration and on the voltage scan rate. It is the area under the peak (in coulombs), which is proportional to the amount deposited in the deposition step. As the scan rate increases the peak becomes narrower and so the peak height will increase. However if the same deposition conditions are used and a fixed scan rate chosen the peak height should be proportional to the concentration of the analyte in the original solution.

Linear sweep stripping voltammetry is almost always carried out with a deposition step in a stirred solution.

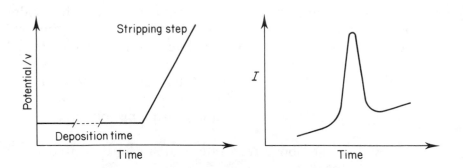

Fig. 4.3a. *The potential signal imposed and the current signal measured in linear sweep stripping voltammetry*

4.3.2. Differential Pulse Stripping Voltammetry

One of the main limitations of dc or linear sweep stripping voltammetry lies in the base line. There is a significant capacitive current as the potential of the electrode is changed. Differential pulse stripping voltammetry offers a much better differentiation of faradaic current signal and noise such as capacitive current.

SAQ 4.3b	What is meant by the terms Faradaic current and capacitive current?

The voltage sweep is identical to that used in conventional differential pulse polarography (Fig. 4.3b). It consists of a linear voltage ramp on which small pulses of 10–100 mV amplitude and about 50 ms duration are imposed at intervals of about 1 s. The current is sampled just before the pulse and almost at the end of the pulse. The final signal is the difference of these two current values.

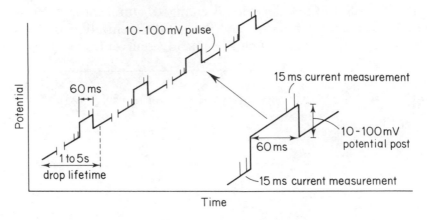

Fig. 4.3b.   *The profile of the potential pulse and current measurement in differential pulse stripping voltammetry*

The two current values measured represent the current at two potential values separated by a few millivolts (the pulse amplitude). At the steeply rising stripping peak this small change in the potential will produce a large change in the electrolysis current whereas the same small change in potential will produce only a very small effect on capacitive current and other sources of noise. As it is the change in the current on either side of the imposition of the pulse which the differential pulse mode records, this mode allows maximum differentiation of the electrolysis current from the background signal. This can be seen in Fig. 4.3c, comparing linear sweep and differential pulse stripping peaks measured under similar conditions.

∏   By analogy with differential pulse polarography, what effect would a change in the pulse amplitude have on the differential pulse stripping peak?

Increasing the pulse amplitude will increase the height of the peak and the sensitivity. However increasing the pulse amplitude will decrease the ability to resolve close lying or overlapping peaks.

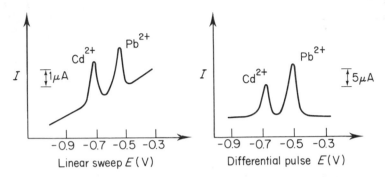

Fig. 4.3c. *A comparison of linear sweep and differential pulse stripping peak for 4 ppb cadmium and lead ions in urine*

The much greater sensitivity of the differential pulse stripping step over the simpler linear sweep stripping step allows the use of much shorter pre-electrolysis or deposition times. One can thus avoid the problems associated with prolonged deposition times, such as loss of proportionality.

The greater sensitivity of the differential pulse technique can sometimes even allow the use of a pre-electrolysis deposition step in an unstirred solution, thus avoiding the problems of the reproducibility of the stirring. For concentrations below 50 ppb stirring is essential but at higher concentrations it can sometimes be avoided.

SAQ 4.3c Why would an unstirred solution offer greater reproducibilty?

SAQ 4.3c

The overall sensitivity of stripping voltammetry depends on the combination of the length of the deposition step and the sensitivity of the stripping step itself. The sensitivity is so high that frequently the overall sensitivity is limited by the purity of the reagents and the sampling and separation techniques. The differential pulse stripping technique is often preferred for its greater reproducibility and reliability. Typical limits of detection for the stripping techniques are 10^{-10}–10^{-9} mol dm^{-3}.

SAQ 4.3d	Convert this range of typical detection limits 10^{-10} to 10^{-9} mol dm^{-3} to ppb for a first period transition metal ($A_r = 50$).

4.4. SUITABLE ANALYTES FOR STRIPPING VOLTAMMETRY

Unfortunately the range of analytes to which stripping voltammetry can be applied is much narrower than for conventional polarography. Several conditions must be met for stripping voltammetry to be possible.

— The product of the deposition step must be insoluble or soluble in the mercury of the electrode. If it were soluble it would simply diffuse away back into the solution and no concentration or build up of the product would occur at the electrode. This rules out the majority of all electro-active organic species.

— If an insoluble product is formed, it must be completely insoluble even at the very high dilutions concerned. It must form a coherent layer on the electrode surface in complete electrical contact. A layer of high electrical resistance must not form.

— The deposited layer must be capable of re-oxidation (or re-reduction) in order to give a stripping peak. That is, it must be possible to reverse the deposition process, although the reaction need not be strictly thermodynamically reversible.

— The deposit must be chemically stable and not be attacked by the solution.

There are many insoluble deposition products which are not suitable for stripping analysis.

4.4.1. Anodic Stripping Voltammetry

This is concerned almost entirely with trace metal analysis, although a few other applications are known. Metal ions lend themselves particularly well to stripping voltammetry. The metal deposited in the deposition step generally dissolves in the mercury drop to form an amalgam. This avoids any problem with the nature of an insoluble deposit.

If more than one metal ion is deposited (at the deposition potential), they will in general appear as separate peaks at different potentials in the stripping step (Fig. 4.3c) and can be determined individually.

However, formation of intermetallic compounds can cause problems. When metals such as copper and zinc are present in solution there is a tendency to form a Zn/Cu intermetallic compound when larger amounts are deposited at a mercury electrode. When an intermetallic compound is formed the stripping peaks for the constituent metals may be shifted, severely depressed, or even be absent altogether. When an alloy is formed at a solid electrode its dissolution potentials, in the stripping step, may be quite different to those of the constituent metals.

SAQ 4.4a	How would you find out if a second metallic ion in the sample would cause interference by forming an intermetallic compound with the analyte?

These interfering effects can be minimised or avoided by reducing the deposition time and the total amount of metal deposited. This would of course mean a loss of sensitivity, but use of differential pulse stripping voltammetry can offset this. The use of a hanging mercury drop electrode with its larger mercury volume offers less intermetallic interference than does the mercury film electrode. Careful choice of a deposition potential can also sometimes prevent the codeposition of metals forming intermetallic compounds.

4.4.2. Cathodic Stripping Voltammetry

The most common species determined by cathodic stripping voltammetry are anions such as halides or sulphide, at a mercury electrode. This involves formation of a film of mercury(I) salts on the electrode in the deposition step. The anodic oxidation process involved in the deposition step is in fact the oxidation of mercury metal to mercury(I) ions. These immediately precipitate insoluble mercury(I) salts with the halide ion etc, on to the surface of the electrode. The anodic deposition potential required depends on the anion concerned. The subsequent cathodic stripping peak for the mercury(I) salt of each anion has at its own individual potential.

Cathodic stripping voltammetry has proved suitable for a number of organic compounds, including drugs and pesticides. These in general contain sulphur and again the deposition step involves anodic (oxidation) formation of an insoluble mercury salt. Clearly this is possible only with a mercury electrode.

A few metal ions such as Mn^{2+} and Pb^{2+} can also be determined by the cathodic stripping of their oxides on carbon or platinum electrodes.

All these cathodic stripping processes involve an initial deposition of an insoluble layer on the surface of the electrode. If too much material is deposited the quality of the stripping peak will deteriorate and the height of the peak will cease to be proportional to the original concentration. The method is highly sensitive but is suited only to these very low levels.

SAQ 4.4b

In one sentence state why stripping voltammetry is the most sensitive of all voltammetric methods?

SAQ 4.4c

In a given electrolyte Fe(III) gives rise to two dc polarographic cathodic waves. On the upper plateau of which wave would the deposition potential be chosen for an anodic striping voltammetric determination of Fe(III)?

SAQ 4.4d

> Could Fe(II) and Fe(III) be differentiated by stripping voltammetry in the electrolyte mentioned in SAQ 4.4c?

SAQ 4.4e

> A sample of saline effluent was divided to give two 20.0 cm^3 aliquots. One of these was subjected to anodic stripping voltammetry. A deposition of 10 minutes at -0.8 V was chosen. A subsequent anodic stripping peak, of 24.6 units in height on the chart paper, was found to correspond to cadmium. To the second 20.0 cm^3 aliquot was added 0.1 cm^3 of a standard solution of 5×10^{-6} mol dm^{-3} of cadmium ions. This gave a new anodic stripping peak for cadmium of 39.8 units under identical conditions. Calculate the concentration of cadmium ions in the effluent in ppb. [$A_r(Cd) = 112.4$]

SAQ 4.4e

SUMMARY AND OBJECTIVES

Summary

Stripping voltammetry is an important, but limited technique, which uses a pre-concentration step to enhance the sensitivity of the voltammetric analysis step. This pre-concentration step is electrochemical using the same instrumentation as the subsequent voltammetric analysis. The simplest polarographic equipment can be used to obtain the highest sensitivity which depends on the length of the pre-concentration step as well as on the sensitivity of the voltammetric stripping step. Since the pre-concentration step requires the production of an insoluble product which can be reproducibly stripped from the electrode surface in the determination step, the use of stripping voltammetry is limited to a few analytes, eg transition metal ions, halides and pseudo-halides.

Objectives

You should now be able to:

- explain the basic principles and practice of stripping voltammetry;

- explain why stripping voltammetry is the most sensitive voltammetric technique and how this sensitivity is achieved;

- describe the types of electrode employed and whether the solution should be stirred or not;

- differentiate between linear sweep and differential pulse stripping procedures and understand the origin of the greater sensitivity of the latter technique;

- recognise the much more limited range of analytes that can be determined by stripping voltammetry in comparison to more direct forms of voltammetry;

- compare the advantages of stripping voltammetry with differential pulse polarography;

- explain what is meant by cathodic and anodic stripping and know in general what kinds of analyte can be determined by cathodic and anodic stripping respectively.

5.   Polarography/Voltammetry and the Analyst

Overview

This part has a very different purpose to the previous parts which have concentrated on the details of polarography itself. An attempt will be made to place polarography and other voltammetric techniques in the full perspective of the analytical environment with real samples. Polarography, spectroscopy etc are not in themselves analytical methods but merely one step in a totally integrated multistep procedure. Other steps used in conjunction with polarography will be discussed.

There is no such thing as the best method for a particular analyte. The role of the sample matrix carrying the analyte is most important. Sometimes polarography will be appropriate, sometimes other techniques such as atomic spectroscopy or hplc. Polarography/voltammetric methods are compared with their rivals in the most general terms. An individual decision must however be made for each individual analytical problem.

5.1.   ANALYTICAL INSTRUMENTATION IN PERSPECTIVE

It is important to realise that polarography, atomic absorption spectroscopy, uv spectroscopy etc are not methods of analysis. They are only the final step in a multi-step process. They represent just the final detection and quantitative determination phase. A typical analytical method might well consist of identification of the analytical information required, sampling and initial treatment of the sample, separation of the analyte, derivitization or pretreatment and final instrumental determination of the analyte, followed by interpretation of the results. The choice of the final determination technique must be made in harmony with all the other steps. Frequently the greatest problems arise in the separation step and the choice of final determination procedure may be of little importance. In other cases it may be vital. Equally the separation step should be chosen to avoid interferences in the chosen determination procedure.

Atomic spectroscopic methods give the total elemental concentration of, for example, chromium while polarographic methods can give the concentrations of the individual oxidation states. The latter is known as speciation. If knowledge of the elemental concentration is sufficient either technique might be chosen. On the other hand chromium(VI) is much more toxic than chromium(III), therefore polarographic analysis generally offers results more indicative of the toxicity of the sample, by selective determination of the two forms. Thus polarography can yield more useful information. However this applies to the determination in the final solution after separation from the original sample. If the separation of the chromium from the sample requires powerful oxidising agents, the original ratio of chromium(III) to chromium(VI) will be disturbed and the ability of polarography to differentiate them will cease to be relevant. In fact it will become a disadvantage. In a quite different case, if the separation step were to include use of an ion exchange resin the anionic chromium(VI) and cationic chromium(III) could be separated and a method, including atomic absorption spectroscopy, could be used for the selective determination of the two forms. Thus this example shows that the relative merits of polarography and other techniques must always be considered in the context of all the steps in the analysis.

The role of the sample matrix is of vital importance. It is quite incorrect to say, for example, that a polarographic method exists for the determination of the steroid, hydrocortisone. One can speak only of a polarographic method for a particular analyte in a specific type of sample matrix. A good example is given by Jacobsen's report of a successful differential pulse polarographic determination of hydrocortisone in certain ointments or creams such as 'Milliderm cream AL' but when an attempt was made to apply this method to another commercial preparation 'Salvizol ointment AL' the method failed completely. The reason for failure lay in the different surface active agents present in the two sample matrices (or excipient as the matrix is often called in pharmaceutical analysis). In Jacobsen's method the separation step was very simple and rapid. The cream or ointment was stirred with alcohol and diluted with a buffer solution. If a more refined separation step could be devised to completely remove all traces of the surface active agent a new polarographic method could be applied to the Salvizol ointment. If an existing polarographic method is to be applied to a new type of sample, the role of the sample matrix must be carefully investigated. If necessary a new method must be developed. This new method might or might not involve the same final polarographic determination step. With some samples a polarographic method might be the most suitable and with others it might not.

In addition to bulk constituents of the sample, trace constituents of the sample might interfere with the determination of a particular trace analyte. The role of possible interferences should always be investigated. Often the choice between a polarographic method and some rival technique may be made on the basis of the role of particular interfering agents in the two techniques. Again the choice should be made in relation to all the steps of the analytical procedure and not just the final instrumental step.

Finally it must always be remembered that almost any instrumental response related to concentration could be used as a method of analysis provided one always had pure solutions. Real samples are almost always very complicated and the role of other constituents must always be considered in any instrumental or other means of analysis.

∏   Summarise the important points of the above section in a single sentence.

In an analytical method the polarographic measurement is only one step in a multistep process in which all the steps must take into account the complex nature of the real sameple.

5.2. THE RELATIVE MERITS OF POLAROGRAPHY AS A DETERMINATION STEP

The most important regions of application of polarography/voltammetry in chemical analysis today are the trace determination of metallic ions particulary in enviromental analysis and the determination of organic species in pharmaceutical, pharmacological and clinical samples. There are of course many others.

5.2.1. Trace Metal Analysis

The main rival to voltammetric methods for trace metal determination is atomic spectroscopy. With a limit of detection about 5×10^{-5} mol dm^{-3} for dc polarography and 1×10^{-5} mol dm^{-3} for sampled dc polarography, these two polarographic methods are similar in sensitivity to atomic absorption spectroscopy using flames. Differential and normal pulse polarography have limits of detection (10^{-8} and 10^{-7} mol dm^{-3} respectively) which are similar to the flameless atomic absorption spectroscopic methods such as the graphite furnace and graphite rod techniques. Stripping voltammetric methods with limits of detection, which can reach as low as 10^{-10} mol dm^{-3} can often be more sensitive than the atomic spectroscopic methods.

Perhaps the most important advantage of the voltammetric techniques over the atomic spectroscopic techniques is the ability of the voltammetric techniques to differentiate between the different oxidation states of the metal, and hence give environmentally more relevant information. As was briefly stated (5.1) this applies only to the final solution used in the instrumental determination step. However it does mean that a simpler separation step can be used prior to the voltammetric procedure and still allow quantitative speciation. In some cases, involving effluent or natural water samples, the separation step can even be eliminated.

Atomic spectroscopic methods have an enormous advantage in their speed. The final determination of a single element by atomic spectroscopy can be made in a minute or less. Most voltammetric determinations take five minutes or more for the potential scan alone. However the situation can be reversed for multi-element determinations. Polarographic or voltammetric methods often allow the simultaneous determination of several species in the same solution. With the atomic spectroscopic methods each element would require a change of lamp and its realignment.

Polarographic methods probably suffer more from interferences of various kinds than do the atomic spectroscopic methods, but the interferences in polarography are probably easier to eliminate by careful choice of supporting electrolyte or at the separation stage. Inter-element effects can be a problem in atomic spectroscopy which is not easy to solve. Overlapping waves in polarography can be resolved for metallic ions, by changing the complexing agents in the supporting electrolyte. A major problem in atomic spectroscopy is refractory oxide formation. In polarographyic analysis interference by surface active agents or unknown complexing agents is the major problem.

An important consideration is the price of the instrumentation. Good polarographic equipment including dc and pulse modes can be bought for £1000 to £2000. Atomic spectroscopic equipment may be £10 000 or more.

SAQ 5.2a	The EEC allowed limit for cadmium ion in drinking water is 0.050 ppb. Briefly consider the general issues on which you might base a decision whether to purchase atomic spectroscopy equipment or polarographic equipment for the determination of cadmium at this level.

SAQ 5.2a

5.2.2. Organic Analysis

The main rivals to voltammetric methods for trace determination of organic species are uv/visible spectrocopy, fluorescence spectroscopy and chromatographic methods such as hplc. With a limit of detection about 5×10^{-5} mol dm^{-3} for dc polarography and 1×10^{-5} mol dm^{-3} for sampled dc polarography, these two polarographic methods are similar in sensitivity to uv/visible spectroscopy. Differential and normal pulse polarography have limits of detection (10^{-8} and 10^{-7} mol dm^{-3} respectively) which are relatively similar to those for fluorescence spectroscopy and hplc techniques. Stripping voltammetric methods can be applied to very few organic species but with limits of detection, which can reach as low as 10^{-8} mol dm^{-3}, it offers a very high sensitivity.

SAQ 5.2b Why can stripping voltammetry be applied to only a very small number of organic compounds?

SAQ 5.2b

The main advantage and disadvantage of polarographic analysis for organic species such as drugs is that only a relatively small fraction of organic compounds are electroactive. Thus there are many drugs which cannot be determined by polarographic means. A clinical or pharmacological laboratory could not rely on polarography alone but would have to have other instrumental methods available. On the other hand the fact that very few compounds present in the sample are electroactive allows a very good degree of selectivity for those drugs and metabolites which are electroactive. For example a simple alkaline or acidic extraction of an electroactive drug from blood plasma with ether will generally not extract much in the way of naturally occuring electroactive interfering species. The electroactive drug would then subsequently give a clearly defined differential pulse polarographic peak. The extract itself will contain a complicated mixture of materials which could be detected by spectroscopic and chromatographic methods giving a complex signal. The electrochemical method will largely ignore the electroinactive constituents.

Π   Why might polarographic analysis of a drug not be carried out directly in blood plasma?

     Blood plasma is a very complex material containing a large number of surface active agents and proteins which also show some electroactive behaviour, but which do not extract into solvents.

An advantage to polarographic and other voltammetric methods of analysis for drugs is that frequently the metabolism of a drug in the body involves a change of its redox properties and hence its polarographic behaviour. A good example is the phenothiazine range of drugs, such as chlorpromazine (Largactil) and promethazine (Phenergan).

Chlorpromazine   S-oxide   N-oxide   S-oxide-N-oxide

$R = -(CH_2)_3 N(CH_3)_2, X = Cl$

Metabolites

Promethazine

$R = -CH_2CH(CH_3)N(CH_3)_2, X = H$

The parent drug itself is not electroactive at the DME but its metabolites the S-oxides and N-oxides are electroactive, by reduction, and can be selectively determined. The parent phenothiazine can be selectively determined by anodic voltammetry at a gold or graphite solid electrode. Individual phenothiazine drugs cannot be differentiated by polarography but it is unlikely that the patient supplying the blood plasma sample would be taking more than one phenothiazine. Frequently electroactive compounds are also strongly therapeutically active or toxic. Since many metabolic mechanisms and polarographic activity both involve changes of oxidation state or redox character, the polarographic analysis can reflect clinical changes. Thus polarographic methods are often suitable for metabolism studies.

Dc polarographic methods have a sensitivity quite suitable for the determination of many drugs in tablets, creams and other pharmaceutical formulations. Frequently it is possible to simply crush and grind the tablet or tablets in water or some suitable solvent such as methanol and then dilute with the chosen supporting electrolyte for direct polarographic analysis. Most excipients used to carry the drug, such as lactose or starch are not electroactive. Fragments of undissolved excipient (tablet matrix) will not interfere with polarographic analysis. These undissolved fragments would interfere badly with a uv spectroscopic method; further spectroscopic methods usually require more complex time-consuming separation or clean-up steps, often involving solvent extraction. The polarographic determination of drugs in ointments or creams can sometimes be just as simple and rapid, but interferences from surface active agents are more of a problem.

∏   Why would undissolved particles from tablets or ointments
        interfere with spectroscopic but not polarographic measure-
        ments?

        Undissolved fragments would cause an apparent light ab-
        sorption due to light scattering by the particles. This will
        apply to particles small enough to pass through filters and
        so elaborate cleaning techniques are sometimes necessary.
        Such particles will have no effect on the diffusion rate of the
        analyte and so will not interfere with a polarographic wave
        height.

Metabolic studies and clinical analyses of drugs in body fluids such
as blood plasma, urine, cerebrospinal fluid require the much higher
sensitivity of normal or differential pulse polarography. A major advantage is that the separation step required for the polarographic
method can often be much simpler than for fluorescence spectroscopy or even hplc. The number of peaks generated by pulse
polarographic analysis will be much less than for hplc with spectroscopic detection. The major disadvantage of the polarographic
analysis is the relatively more restricted range of drugs to which
it can be applied and the dangers of interference from surface active agents. The expression 'relatively restricted range', in fact, still
leaves a very large range of organic species open to polarographic
analysis. A further advantage is the relatively low cost of polarographic equipment; the cost of hplc equipment can be as much as
ten times greater. Times per analysis are relatively similar.

5.3.   SEPARATION STEPS EMPLOYED WITH ORGANIC
         POLAROGRAPHIC ANALYSIS IN BIOLOGICAL
         FLUIDS

Direct polarographic analysis in biological fluids is rare due to the
interferences from surface active agents. Further proteins themselves can show a complex electroactive character. Thus various
separation steps are usually carried out prior to the polarographic
determination. These steps can minimise the interference and can
also be used to enhance the selectivity of the method. Many common separation steps are used in combination with polarographic
analysis.

(*a*) Ion exchange chromatography. This has been used to remove the anionic electroactive interfering species from urine samples prior to the determination of various drugs. Naturally occuring electroactive species in urine samples are mostly organic acids.

(*b*) Thin layer and paper chromatography. Polarographic analysis can be carried out after extraction of the analyte from the spot with a suitable solvent. The precision is not high but can allow good resolution of compounds of similar half wave or peak potentials.

(*c*) Adsorption, gel filtration and other forms of chromatography. These also have been employed by various workers.

(*d*) hplc. Although this can be employed with the DME it is more usual to employ small tubular graphite electrodes or wall jet electrodes in the eluate flow from the hplc column. It is usual to speak of hplc with voltammetric detection. This combination is very powerful, as it combines the great power of separation of the hplc technique with the selectivity of voltammetry. In quite a few cases the best possible hplc separation yields many overlapping peaks which are not adequately resolved and confusing to interpret. By using voltammetric detection only the minority of electroactive species are detected and a much simpler instrumental response is obtained with clearly resolved peaks. Complete current-potential curves could be obtained at regular intervals using a fast potential scan but usually the current is recorded as a function of retention time at a chosen fixed potential.

(*e*) Dialysis. This can be used to separate small electroactive molecules of interest from high molecular mass proteins which themselves are electroactive.

(*f*) Solvent extraction. This is the most commonly used separation technique since it can remove many naturally occuring electroactive interfering species and, with care, offer an enhanced degree of resolution between closely related compounds such as the metabolites of a drug. It can also serve as a preconcentration step.

There are two main variables in solvent extraction – the polarity of the solvent and the pH of the sample solution. Generally speaking the less polar the solvent the less naturally occurring electroactive interfering species will be extracted from biological fluids. Relatively 'clean' polarographic blanks can be obtained following solvent extraction from, for example blood plasma, at a wide range of pH values. A drug and its metabolites will differ in their acid-base chemistry. Generally only the neutral forms can be extracted into organic solvents. By careful choice of pH and the polarity of the solvent, various metabolites can be selectively extracted in a series of extraction steps. Thus the separation step is used to increase the selectivity of the subsequent polarographic determination. The organic extracts are usually evaporated to dryness and the residue containing the analyte is taken up in the buffered supporting electrolyte for the polarographic step. A back extraction into a buffered aqueous solution is also possible.

| **SAQ 5.3a** | What are the three most important roles played by the separation step? |

5.3.1. A Complete Analytical Method Using Polarography in Clinical Analysis

A good example of a solvent extraction scheme combined with polarographic determination is given by the determination of chlorpromazine (Largactil) and its three metabolites (the S-oxide, the N-oxide and the S-oxide-N-oxide) in blood plasma. The details of this method are not important and you should make no effort to remember them. However it does serve as a good illustration of a complete polarographic method.

It has been mentioned in a previous section that polarography can distinguish between the electroinactive (at the DME) chlorpromazine and its electroactive metabolites. However the three metabolites are all reduced (back to the parent drug) at similar potentials and their polarographic waves or peaks cannot be resolved. The S-oxide-N-oxide produces twice the wave height since it has two electroactive groups.

An extraction scheme has been devised by Beckett (Fig. 5.3a). The cationic dye methyl orange is used to release the drug and metabolites from protein binding to allow extraction. The chlorpromazine and its N-oxide are extracted into benzene/dichloroethane (95:5) v/v at pH 4, leaving the S-oxide and S-oxide-N-oxide in the aqueous sample solution. The latter are then subsequently extracted by making the solution alkaline and using the more polar solvent dichloroethane in a second extraction. The S-oxide and the S-oxide-N-oxide are determined polarographically together. They are differentiated by the reduction in wave height when the S-oxide-N-oxide is reduced to the S-oxide with SO_2. The chlorpromazine and its N-oxide extracted in the first extraction are further separated in a series of additional extraction steps. The N-oxide is finally determined by direct differential pulse polarography. The chlorpromazine can be determined by voltammetry at a gold electrode (electrooxidation) or at the DME after oxidation with bromine water to the S-oxide.

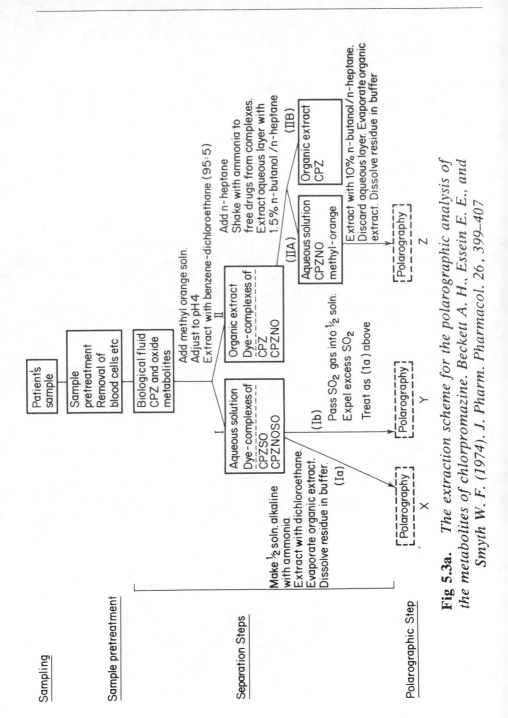

Fig 5.3a. *The extraction scheme for the polarographic analysis of the metabolites of chlorpromazine. Beckett A. H., Essein E. E., and Smyth W. F. (1974). J. Pharm. Pharmacol. 26, 399–407*

CPZ
Chlorpromazine

CPZSO
S-oxide

CPZNO
N-oxide

CPZNOSO
N-oxide-S-oxide

$$R = -(CH_2)_3 N (CH_3)_2$$

Wave height of X is proportional to CPZSO + 2 × CPZNOSO
Wave height of Y is proportional to CPZSO + 1 × CPZNOSO
(X and Y come from half the total sample)
Wave height of Z is proportional to CPZNO

NB CPZNOSO gives twice the wavelength of CPZSO or CPZNO

It is important to realise that this analytical method begins as the needle leaves the arm of the patient. It is vital, for example, to separate the red blood cells from the sample as soon as possible, because the breakdown of haemoglobin would introduce electroactive iron species which might be carried through to the final polarographic steps. The method is an integrated process in which the sampling and separation steps are harmonized with the final polarographic steps. The above example takes advantage of the great sensitivity of differential pulse polarography, but the selectivity and the metabolic speciation, in this case, depend on the separation steps.

SAQ 5.3b   In the above example the selectivity between the
species of interest is obtained in the solvent extraction steps. What might be the advantage in
using polarography for the final determination
step?

5.4. THE USE OF DERIVITIZATION IN
POLAROGRAPHIC ANALYSIS

Polarography requires an electroactive species. However compounds which themselves are not electroactive can be be determined
using polarography by chemically changing them into electroactive
species. With organic analytes this usually involves the introduction
of an electroactive substituent. Such a process is known as derivitization. Some common derivitization reactions are listed.

— Nitration. Compounds containing aromatic nucleii can often be readily be nitrated. The introduced nitro group gives a well defined analytically useful wave or peak.

— Reaction with sodium nitrite. This reaction gives rise to a variety of electroactive products depending on the compound concerned and the reaction conditions. This type of reaction is most often applied to phenols and amines.

— N-oxidation and S-oxidation. This can be used for heterocyclic nitrogen and sulphur compounds. Oxidising agents could include hydrogen peroxide, bromine and chloroperbenzoic acid. A good example would be the oxidation of electroinactive chlorpromazine to its electroactive S-oxide.

— Formation of semicarbazone and hydrazone derivatives. This can be used for electroinactive ketones and aldehydes. The derivatives are electroactive.

— Hydrolysis and other forms of degradation. Degradation of some species creates electroactive products. Acceleration of these reactions to completion could be used to allow polarographic analysis.

Generally derivitization is not very selective as other solution components may also be involved producing electroactive forms. The precision of the method will also be reduced. It is a technique of last resort in most cases.

5.5. SURFACE ACTIVE AGENTS AND POLAROGRAPHIC ANALYSIS

The analytical chemist using polarography must always be aware of the presence of and effect of surface active agents. These can have a drastic effect on the polarographic behaviour of the analyte. The half wave potential can be shifted sometimes by hundreds of millivolts. The wave shape can change and the height may or may not be depressed. Sometimes the wave is split into new waves or is so badly formed as to be useless. On occasions the wave may be completely suppressed. Certainly a calibration made with pure solutions will no longer be valid.

The situation, however, is not hopeless. There are various ways of coping with samples containing surface active agents.

In many cases, beyond a certain concentration of a particular surfactant the effect on the polarographic wave or peak becomes constant. Often if the concentration of the surfactant can be kept constant, its distortion of the polarographic behaviour can be kept constant. For example there are many drugs which can be determined in specific ointments by simply stirring the ointment with methanol and diluting the extract with the chosen buffered supporting electrolyte. The resultant polarographic wave or peak can be highly reproducible, with the wave or peak height proportional to the concentration of the analyte, provided the amount of the ointment base, the excipient, is kept constant. The position and height of the wave or peak will be quite different to that of the same amount of the drug in a pure solution in the absence of the ointment base. However if the calibration is made in the presence of the same amount of excipient an accurate analytical determination can be made.

When the analyte is present at higher concentrations, for example vitamin C (ascorbic acid) in orange juice, it is possible to minimise surface active interference by a very large dilution of the sample. The analyte can then be determined using the high sensitivity of differential pulse polarography. This also allows use of a very small sample volume. Vitamin C can be determined in 25 μl samples of fruit juice by dilution with, for example, 5 cm^3 of a pH 3 acetate buffer solution. The differential pulse polarographic peak for the ascorbic acid is well formed and analytically useful. Attempts to use dc polarography to accurately determine the ascorbic acid (about 10^{-3} mol dm^{-3}) directly in the fruit juice or with a slight dilution would probably fail due to surfactants present. Such a dilution technique will not be of use with stronger surfactants.

If the amount and type of surface active agent present cannot be controlled or the interference too drastic or unreproducible, steps must be taken to separate the analyte from such agents or some technique other than polarography used.

Π How might one know if an unexpected surfactant was interfering with the determination of an individual sample?

It is hard to be absolutely sure this has not happened. However, the wave or peak for individual samples should be ex-
amined to check that the half wave (peak) potential and wave
(peak) shape has not altered from those of the standards.
Generally when the wave height or peak height is depressed
by surfactants the potential of the peak/wave is shifted and
the width altered. The use of the standard addition technique avoids this problem by producing the calibration in
the individual sample itself with any surfactant levels present
unaltered.

5.6. CALIBRATION METHODS

In the development of a polarographic method of analysis, the first
step is to establish that the analyte gives rise to a diffusion controlled
wave (peak) whose height is proportional to the concentration of analyte, and, where possible, independent of pH. Then the nature of
the sample matrix and possible interfering species should be investigated and the sampling and separation steps chosen accordingly.
The final analytical signal in polarography is, of course, the height
of the wave or peak (or area under the peak). The final development step is to choose a method of calibration for the wave or peak
height against concentration.

The two most important techniques are the construction of a calibration plot using standards and the method of standard addition.

5.6.1. Calibration Plots

The preparation of a calibration plot using pure solutions, of the
pure analyte in the chosen buffer solution, is of little practical use in
the analytical context. It is of use in establishing whether or not the
wave is diffusion-controlled or not. It is also of use in investigating
the extraction efficiency of the preceding separation step, or the
effect of surface active agents in the sample on the wave or peak
height.

The analytical calibration plot must always be prepared with standard solutions, which are as close as possible in composition to the real samples being investigated. For example when determining a drug in an ointment or cream, it is important that the standards should be prepared containing the same amount of ointment or cream base. Sometimes the standard solutions, for calibration purposes, are simply prepared to resemble the final sample solution subjected to polarography after the separation steps. However it is best to subject the standards to all the steps of the analytical method. This is vital if the preceding extraction method chosen is not 100% efficient.

It is particularly important to ensure that the amount and character of any surface active components in the sample solution are as exactly similar as possible in the standards chosen for calibration of the wave height.

A calibration plot should contain at least six points and cover a range somewhat wider than the expected sample range. Several subsamples should be obtained from the original sample and each subjected to the complete analytical method. An average result for the concentration of the analyte can then be obtained by reference to the calibration plot. This presents the most precise method of interpretation of polarographic analytical results.

The main weakness of such a calibration plot lies in the difficulty of preparing standards and standard solutions to resemble 'real' samples. Human blood plasma samples are surprisingly similar and normal plasma samples spiked with the analyte can often be used to prepare standards for the clinical analysis of the drug by a polarographic analysis. On the other hand natural water samples can vary enormously in composition and preparation of representative standards can often be a major problem.

| **SAQ 5.6a** | What is the difference between accuracy and precision? |

SAQ 5.6a

SAQ 5.6b   The preparation of a calibration plot and of a
sample measurement involves the running of a
'blank'. What is a 'blank'?

5.6.2. The Method of Standard Addition

This method is generally significantly less precise than the use of a calibration plot, but may often be more accurate if the composition of the sample matrix is unknown or too variable. It is in practice a calibration plot prepared in the individual sample solution itself and therefore under the exact, if unknown, conditions of the sample and with the same concentration of surface active agents etc.

The sample solution (eg 5 cm^3) is taken and the polarographic wave or peak obtained. A small volume (eg 5 μl) of a concentrated solution of the analyte of exactly known concentration is added to the sample solution and the new increased polarographic wave or peak height is obtained. The volume added is chosen so as not to appreciably dilute or alter the sample solution except as regards the analyte concentration. Thus any interference or suppression of the wave or peak height from surface active agents etc will remain constant. The standard addition can be repeated several times.

A plot is finally made of wave or peak height against the total quantity of analyte added (Fig. 5.6a). The original wave or peak height lies on wave height axis at zero addition. The plot should be a straight line. This line is extended until it crosses the quantity axis. The negative intercept on this axis gives the original quantity of analyte present in the sample solution.

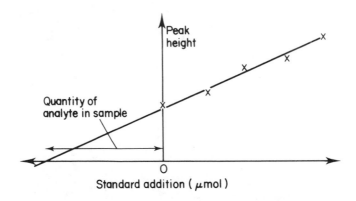

Fig. 5.6a. *A typical standard addition plot*

The most precise results can be obtained by using several standard additions and by repeating the entire procedure with several sub-samples of the original sample. A blank measurement should also be taken, to determine any traces of analyte (or other electrochemically indistinguishable material) present in the reagents used for the supporting electrolyte etc. Sometimes if a lower precision is acceptable and speed of analysis is important, only one standard addition is made and the concentration of analyte obtained by simple calculation from the increase in the wave height (See 1.6.3).

The great advantage of the standard addition technique is that any suppression of the wave or peak height or change in the shape of the wave or peak due to surface active agents will usually be identical for each peak measured, as the concentration of this interfering species is not altered. If the degree of suppression depends on the analyte concentration, the standard addition plot will not be linear and the method will have to be abandoned.

If the character of the sample matrix is reproducible and known, then the method of the calibration plot is best. If the character of the sample matrix is unknown or variable the standard addition technique is to be recommended.

SAQ 5.6c How in fact would you determine which of the two calibration methods to use for a particular analytical problem?

SAQ 5.6c

SUMMARY AND OBJECTIVES

Summary

Voltammetry or polarography, as with any other analytical technique, is only employed as one part of a multistage method in which the electrochemical or polarographic part is simply the final step. The whole analytical procedure must be designed as a coherent whole. Voltammetric/polarographic methods have a number of advantages and disadvantages in comparison with such methods as atomic spectroscopy (for metals) and hplc chromatography (organic analytes). However there is never a 'best method' but just a compromise between the factors involved in each individual case. The role of the sample matrix is an important consideration in the choice of method.

Objectives

You should now be able to:

- give a general account of the philosophy of voltammetric analysis as a tool;

- recognise that voltammetric or polarographic analysis is a multi-stage technique in which the electrochemical part is simply the final step (the whole analytical procedure must be designed as a coherent whole);

- describe the general advantages and disadvantages of voltammetric/polarographic methods in comparison with rival methods (eg. AAS and hplc);

- recognise that there is not a best method but just a compromise between many factors in each individual case;

- explain the role that the sample matrix may play and the major hazard that adsorption effects represent to reliable analysis;

- describe the two most common calibration techniques and be able to choose intelligently between them.

Self Assessment
Questions and Responses

SAQ 1.1a

> Explain what feature of polarography sets it apart from other voltammetric techniques.

Response

It is important to realise that an analytical technique should not be given the name polarography unless the working electrode is specifically a dropping mercury electrode (DME). Polarography is voltammetry at controlled potential using a DME as the working electrode.

****

SAQ 1.2a

> Sketch and label a circuit for a potentiostatically controlled three-electrode cell and explain how potentiostatic control is achieved and maintained. Why are three electrodes necessary? What problems arise when non-aqueous solutions are used?

Response

A suitable circuit is that shown in Fig 1.2d. Potentiostatic control is achieved by means of the auxiliary circuit shown in Fig. 1.2b and 1.2d. The reading on the high impedance voltmeter (P) has the significance of:

$$P = \mid E_{WE} - E_{RE} \mid = E_{WE}(SCE),$$

if a saturated calomel electrode (SCE) is used as the reference electrode. The required value of E_{WE} (SCE) is fed into the potentiostat memory and the potentiostat compares the required value with the measured value. Any difference is an error signal. The potentiostat causes the dc supply to alter in such a way (altering the cell current as a consequence) that the error signal is reduced to zero. When this is achieved (very rapidly) we have potentiostatic control. Control is maintained by the continuous comparing of required and measured signals and the consequent adjustment of the dc supply.

Three electrodes are necessary to overcome the fact that the potentials of most electrodes change when continuous current is passed (polarisation). In a two-electrode cell the measured voltage is the difference in potential between two changing individual electrode potentials. In these circumstances it is impossible to calculate the potential of any one electrode. Non-aqueous solvents cause problems in terms of solubility of the analyte and supporting electrolyte but principally problems arise because the resistance of the solution rises. This causes loss of potentiostatic control. The reference electrode may also become unstable.

****

SAQ 1.2b   Sketch the appearance of a cell suitable for dc
                      polarography analysis and comment on the fea-
                      tures.

Response

Your sketch should have resembled that in Fig. 1.2g and the main
features incorporated should be; (i) thermostatting, (ii) filling and
emptying the cell *in situ*, (iii) purging with gas, (iv) gas venting, (v)
minimum volume, (vi) three electrodes.

****

SAQ 1.3a   Summarise the factors that determine the volt-
                      age window for dc polarography in aqueous so-
                      lution.

Response

The factors are:

(i)   the cathodic voltage limit for the reduction of hydronium ions
         on mercury which is always more negative than -0.8 V (SCE),
         the exact value depending upon the pH;

(ii)   the cathodic voltage limit for the reduction of the cation of the
         supporting electrolyte which for K^+ ions in aqueous solution
         is about -2 V (SCE);

(iii) the anodic voltage limit for the oxidation of mercury which oc-
         curs at potentials more positive than about -0.1 V (SCE), the

exact value depending upon other species present in solution, but never >0.2 V (SCE);

(*iv*) complete removal of oxygen from the solution. If present, oxygen is reduced in two waves which obscure the window.

****

SAQ 1.3b Sketch and describe the construction of a typical DME and relate the quantities *m, h* and *t*.

Response

Your sketch should resemble Fig. 1.3a, the main features are the height of the column of mercury (*h*) in the range 20–100 cm, the capillary of length 10–20 cm and diameter of about 50 μm, a drop of maximum diameter of about 1 mm, a drop rate (*m*) of about 0.5 to 2 mg s^{-1} and a drop time (*t*) of 0.5–5 s. The relationships between *m, h*, and *t* are:

$$t \propto m^{-1} \propto h^{-1}$$

Modern instrumentation incorporates a tapping mechanism which knocks drops from the capillary at a pre-determined interval in the range given above.

****

SAQ 1.3c Explain the origin of a capacitive current in dc polarography and state the time dependence of this quantity.

Response

There is always a characteristic potential for any electrode/solution interface where there is no net charge on the electrode. This is the potential of zero charge. (E_{PZC}) which for Hg/H_2O has a value of about -0.5 V (SCE). At any other potential, charge develops on the electrode and there is an analogy with a parallel plate condenser across the electrical double layer formed at the electrode. A capacitance is created, C_{DL}, and a capacitive current, I_{CP} is given by:

$$I_{CP} \approx C_{DL}E(PZC)\,\delta A/\delta t$$

Treating the Hg drop as a sphere,

$$I_{CP} \propto t^{-\frac{1}{3}}$$

This time dependence shows a rapid decay during the drop lifetime. The capacitive current increases (in positive sense) as the potential moves negative to E_{PZC}.

****

SAQ 1.3d	Sketch the appearance of current maxima in dc polarography and explain how these unwanted signals can be avoided.

Response

Your sketch should resemble Fig. 1.3f. There are two kinds of maxima and they are both due, in differing ways, to an enhancement of the required diffusion-controlled mass transport mechanism by convective movements at the Hg/H_2O interface. Both types of maxima may be prevented by the addition to the solution of a small amount of surfactant, eg Triton X-100.

****

SAQ 1.4a	Select a suitable solvent/supporting electrolyte system for the following applications using the dc polarography technique.

> (i) Determination of Cd(II) in an approximately 0.001 mol dm^{-3} aqueous solution of Cd(NO$_3$)$_2$.
>
> (ii) Determination of nitrobenzene at the 10 ppm level in methanol.
>
> (iii) Determination of Pb(II) at a concentration of about 10^{-6} mol dm^{-3} in an aqueous solution.

Response

(i) There is no need to do anything other than use the system H$_2$O/0.10 mol dm^{-3} KCl. This will give a window of about 0 to -2 V (KCl) and you would expect the wave for Cd(II) to have $E_{\frac{1}{2}} \approx -0.6$ V (SCE).

(ii) Here it would be necessary to give yourself as wide a window as possible because you have no information on the value of $E_{1/2}$ for nitrobenzene. Use say 0.10 mol dm^{-3} (C$_2$H$_5$)$_4$NBF$_4$ as supporting electrolyte. this should, with methanol as solvent, give a window of at least $+0.2$ to -2.5 V (SCE).

(iii) This is a trap. You should have said that 10^{-6} mol dm^{-3} is below the detection limit for dc polarography. More advanced polarographic methods will detect this level of concentration easily. There are also general techniques available for concentrating metallic cation solutions.

****

SAQ 1.4b	What are the two roles of the supporting electrolyte in dc polarography?

Calculate the relative magnitudes of the migration current and the diffusion current for the zinc cation in an aqueous solution at 25 °C containing 5×10^{-4} mol dm^{-3} Zn(NO$_3$)$_2$ and 0.1 mol dm^{-3} KCl.

Mobility values /m^2 V^{-1} s^{-1}:

NO$_3^-$ 7.4 $\times$ 10^{-8}; K$^+$ 7.6 $\times$ 10^{-8};

Cl$^-$ 7.9 $\times$ 10^{-8}; Zn^{2+} 5.5 $\times$ 10^{-8}.

Response

The two roles are:

(i) to make the solution sufficiently conducting by ensuring a low electrical resistance. This is particularly important when non-aqueous solvents are used;

(ii) in dc polarography the theoretical basis of the technique is founded on the assumption that mass transport of analyte to the electrode is diffusion-controlled. This is ensured by using a supporting electrolyte in excess concentration so that the ions of the supporting electrolyte carry most of the migration current.

Answer to the calculation is $I_m/I_d = 0.0035$

If you did not obtain this answer then you should read again 1.4.2 and try the calculation again before reading the solution that follows.

$$t(\text{Zn}^{2+}) = (2 \times 5.5 \times 10^{-8} \times 5 \times 10^{-4})/\text{Denominator}.$$

Denominator $= (2 \times 5.5 \times 10^{-8} \times 5 \times 10^{-4})$
$+ (1 \times 7.4 \times 10^{-8} \times 10^{-3})$
$+ (1 \times 7.6 \times 10^{-8} \times 0.1)$
$+ (1 \times 7.9 \times 10^{-8} \times 0.1)$

$= 1.56 \times 10^{-8}$

Thus $t(Zn^{2+}) = 55 \times 10^{-12}/1.56 \times 10^{-8}$

$= 3.52 \times 10^{-3}$

A fraction 0.0035 of the total cathodic current is due to the migration of Zn(II) ions to the cathode.

$\therefore$ $I_m \propto 0.035$ units, if total current is say 10 units.

$\therefore$ $I_d \propto 9.965$ units and,

$I_m/I_d = 0.0035$

We see that under these conditions the transport of Zn(II) to the electrode is under diffusion control.

****

SAQ 1.5a   State the main contributions to the Faradaic current in an electrolysis cell. How does diffusion control manifest itself in the current/working electrode potential relationship?

Response

The Faradaic current is a measure of the rate of all the electro-chemical reactions occurring at the working electrode. We say that there is 100% Faradaic efficiency if the only reaction occurring is the required one. The main contributions to the Faradaic current are:

— the rate of the overall electron transfer process at the electrode surface, and

— the rate of mass transport of the electroactive species through the solution to the electrode.

Diffusion control manifests itself in the existence of a maximum (limiting) current which in dc polarography is called the diffusion current (I_d).

****

SAQ 1.5b   Show by a sketch the effect of slow and fast electron transfer and slow and fast mass transfer upon the shape of I/E_{WE} curves.

Response

The required sketch should resemble that in Fig. 1.5n.

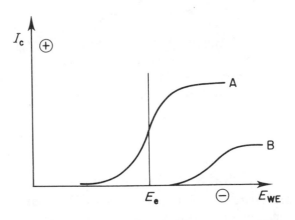

Fig 1.5n.   *I/E_{we} curves*

Curve A shows fast electron transfer and mass transport, curve B slow electron transfer and mass transport. There is a significant change in the position and height of the wave. Note the position of E_e, the reversible potential.

****

SAQ 1.5c

State the general form of the Ilkovic equation for a reduction process and explain the terms used. From this equation derive the form of the Heyrovsky-Ilkovic equation that applies to the reaction:

$$\text{red} \rightarrow \text{ox} + n\text{e}$$

where both reduced and oxidised species are soluble, and only the reduced species is present in the solution. Draw the wave obtained in this situation.

Response

The general form of the Ilkovic equation is:

$$I = 708 \, nFD_{\text{ox}}^{\frac{1}{2}} \, m^{\frac{2}{3}} \, t^{\frac{1}{6}} \left(c^{\text{ox}} - c_0^{\text{ox}} \right),$$

where n is the number of electrons transferred, $D/m^2 \, s^{-1}$ the diffusion coefficient; $m/\text{kg} \, s^{-1}$ the rate of flow of Hg; t/s the drop time, $c^{\text{ox}}/\text{mol m}^{-3}$ the concentration of analyte in the bulk solution, $c_0^{\text{ox}}/\text{mol m}^{-3}$ the concentration of analyte at the electrode surface.

For an oxidation process,

$$I = 708\ nFD_{red}^{\frac{1}{2}} m^{\frac{2}{3}}\ t^{\frac{1}{6}} \left(c_0^{red} - c^{red} \right)$$

$$\therefore \quad I = k(0 - c_0^{ox}); \quad I = k' \left(c_0^{red} - c^{red} \right)$$

Note that $c^{ox} = 0$, because only reduced species is present.

From Nernst equation:

$$E = E^{\ominus} - \frac{RT}{nF} \ln \frac{c_0^{red}}{c_0^{ox}}$$

$$= E^{\ominus} - \frac{RT}{nF} \ln \left\{ \frac{(I + k'c^{red})/k'}{(-I/k)} \right\}$$

In this case

$$I_d = -k'\ c^{red}; \quad c_0^{red} = 0.$$

$$\therefore \quad E = E^{\ominus} - \frac{RT}{nF} \ln \left\{ \frac{(I - I_d)/k'}{(-I/k)} \right\}$$

$$\therefore \quad E = E^{\ominus} - \frac{RT}{nF} \ln \frac{D_{red}^{\frac{1}{2}}}{D_{ox}^{\frac{1}{2}}} - \frac{RT}{nF} \ln \frac{I_d - I}{I}$$

$$= E_{\frac{1}{2}} - \frac{RT}{nF} \ln \left\{ \frac{I_d - I}{I} \right\}$$

Compare with Eq. 1.51 (p65).

The wave obtained would resemble that shown in Fig. 1.5o.

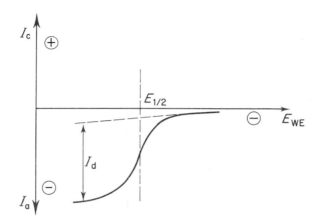

Fig 1.5o. *Polarographic wave for an anodic process*

****

SAQ 1.5d

What is the main factor that determines the limit of detection in dc polarography? Sketch the appearance of the polarograms of a typical supporting electrolyte and a typical analyte solution when the analyte concentration is $>10^{-3}$ mol dm^{-3} (including the shape of the oscillations in current due to drop size).

Response

It is the capacitive current component of the total cell current which is the main factor in setting the detection limit in dc polarography. When the concentration of analyte falls to $< 10^{-4}$ mol dm^{-3} the Faradaic current falls to the level of the capacitive current and the Faradaic signal becomes lost in the background. The polarograms required are in the text. You are being asked to bring together Fig. 1.3e. and Fig. 1.5f. The shape of the individual oscillations are magnified in Fig. 1.3d. and 1.5d.

As the concentration falls to about 10^{-4} mol dm^{-3} the shapes of the curves in Fig. 1.3e. and 1.5f. will merge and the individual oscillations will as a consequence resemble Fig. 1.5e.

$$*********************************$$

SAQ 1.5e	Use a graphical method and the Meites method to determine whether or not the following data relates to a reversible reaction. What is the significance of $E_{\frac{1}{2}}$ for a reversible reaction; is this significance valid for this reaction?

E_{WE} (SCE)/V: -0.419 -0.451 -0.491 -0.519 -0.561

$I/\mu A$: 0.31   0.62   1.24   1.86   2.48

The reaction is

$$ox + 2e \rightarrow red$$

$I_d = 3.10\ \mu A$; temperature 25 °C.

Response

The answer is that the reaction is not reversible. The graphical method is to plot log $I/(I_d - I)$ against E_{WE}, (Fig. 1.5p.)

log $I/(I_d - I)$: -0.955 -0.602 -0.176 0.176   0.602

E_{WE}/V : -0.419 -0.451 -0.491 -0.519 -0.516

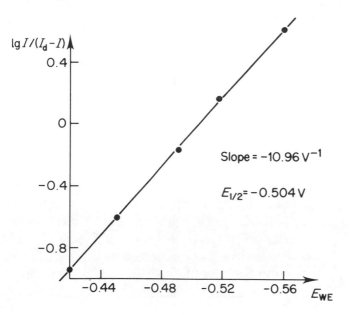

Fig. 1.5p. *Test for reversibility*

If the reaction is reversible then the slope should be $-nF/2.303\ RT = -2/59.1 = -0.0338\ \text{mV}^{-1}$. We see that for the above data the slope is $-0.0110\ \text{mV}^{-1}$; numerically much too small. Hence the reaction is irreversible.

The Meites test is normally applied to the experimental polarogram but we can calculate $E_{\frac{1}{4}}$ and $E_{\frac{3}{4}}$ from Fig. 1.5p. $E_{\frac{1}{4}}$, ie potential when $I = \dfrac{1}{4}\ I_{\text{d}}$, is -0.536 V; $E_{\frac{3}{4}}$, ie potential when $I = \dfrac{3}{4}\ I_{\text{d}}$, is -0.494 V.

$$\therefore \quad E_{\frac{1}{4}} - E_{\frac{3}{4}} = -0.042\ \text{V} = -42\ \text{mV}$$

For a reversible reaction the value is $\left(\dfrac{-0.0564}{n}\right)\ \text{V} = -0.0282\ \text{V} = -28.2\ \text{mV}$. These values differ greatly so the reaction is irreversible.

For a reversible reaction,

$$E_{\frac{1}{2}} = E^{\ominus} - \frac{RT}{nF} \, ln \left(\frac{\gamma_0^{red} D_{ox}^{\frac{1}{2}}}{\gamma_0^{ox} D_{red}^{\frac{1}{2}}} \right)$$

Although a value for $E_{\frac{1}{2}}$ has been calculated in Fig. 1.5p, -0.504 V (SCE), this value is now a function of k^0 and t as well as γ_0 and D.

****

SAQ 1.5f

> If you had a mixture of Sn(IV)/Sn(II) ions in solution and the concentration ratio of Sn(IV):Sn(II) was 3:1 sketch the appearance of the polarogram expected assuming that the reaction $Sn^{4+} + 2e \rightleftharpoons Sn^{2+}$ is reversible.

Response

Your answer should be derived from knowledge embodied in Fig. 1.5l., and should resemble Fig. 1.5q.

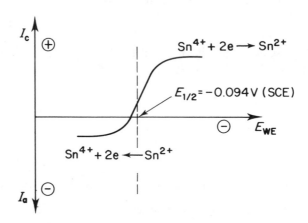

Fig. 1.5q.   *Mixed wave with excess of oxidised species*

Note that the value of the height of the cathodic wave is greater than that for the anodic wave as the concentration of Sn(IV) is greater than the concentration of Sn(II).

Note also that the mixed wave is still centred around $E_{\frac{1}{2}}$ since this is a reversible reaction. The value -0.094 V (SCE) is derived from Fig. 1.2c. after correction to the SCE scale.

****

SAQ 1.6a

> Describe a typical dc polarography experiment designed to obtain the polarogram of a mixture of Pb(II) and Zn(II) ions in their aqueous solutions as nitrates at 25 °C and both at concentrations of about 10^{-4} mol dm^{-3}. Sketch the expected polarogram and show how you would determine the value of $E_{\frac{1}{2}}$ for each analyte.

Response

This is asking you to recall the contents of 1.6.1 and 1.6.2. You should have included the following points in your description.

(*i*)   A circuit similar to that in Fig. 1.2d., using three electrodes and potentiostatic control.

(*ii*)   A 3-electrode cell similar to that in Fig. 1.2g. using a DME (WE), Pt (SE), SCE (RE). Check that the DME is delivering Hg at the required rate. Adjust the Hg column height (h) if necessary or alter the time interval of the mechanical tapper.

(*iii*) Turn on the thermostat water circulation and adjust the thermostat to 25.0 ± 0.2 °C.

(*iv*) Check that the nitrogen supply for deoxygenation is delivering nitrogen at a suitable rate.

(*v*) Make an aqueous solution of about 10^{-4} mol dm^{-3} Pb(NO$_3$)$_2$ and Zn(NO$_3$)$_2$ and 0.1 mol dm^{-3} KCl. The latter is the supporting electrolyte.

(*vi*) Introduce this solution into the cell. Make sure that all three electrodes are in the solution and add a small amount of a dilute aqueous solution of Triton X-100 as a maximum suppressor.

(*vii*) Deoxygenate the solution for about 10 minutes then bathe the surface of the solution with nitrogen (Fig. 1.2h.).

(*viii*) Wait for the solution to become quiescent and also allow time for the solution to attain thermostat temperature.

(*ix*) Set the potentiostat controls to scan 0 to -2 V (SCE) at say 2 mV s^{-1} and set the X-Y recorder to initial conditions. The cell current is fed to the Y-axis and the value of E_{WE} to the X-axis.

(*x*) Complete the circuit and start the scan. The result should have the appearance shown in (Fig. 1.6i.)

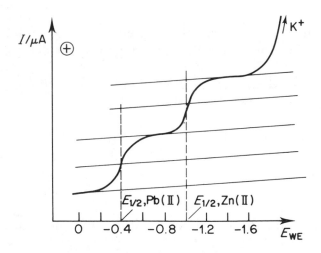

Fig. 1.6i.   *Determination of $E_{\frac{1}{2}}$ for a 2-component system*

The construction lines shown are those described in 1.6.2 and shown in Fig. 1.6b. You may have decided to sketch in the oscillations due to the varying size of the drop. These should appear as in Fig. 1.5f. The values of $E_{\frac{1}{2}}$ shown are given in Fig. 1.6c.

****

SAQ 1.6b

Consult the data in Fig. 1.6c and comment on the feasibility of analysing for Pb(II) and Tl(I) ions in the same solution at 25 °C. What difference would it make if Pb(II) were monovalent or if the concentrations of the two ions differed greatly?

Response

In order to resolve neighouring waves satisfactorily the difference in half-wave potentials should be,

$$\Delta E_{\frac{1}{2}} > \frac{300}{n} \ mV$$

In this case we could take n to be 3/2, $\therefore \Delta E_{\frac{1}{2}} > 200$ mV is sufficient. From Fig. 1.6c. we see that $\Delta E_{\frac{1}{2}}$ is $-0.40 - (0.46) = 0.06$ V $= 60$ mV. Therefore these two waves would overlap making accurate analysis impossible.

If Pb(II) were monovalent matters would be worse because now $\Delta E_{\frac{1}{2}} > 300$ mV would be required. This is because the wave rises more steeply as n increases (Fig. 1.6f.). The problem of resolution is made much worse if the second wave, here Tl(I), is much larger than the first wave which would be true if the Tl (I) concentration were much greater than that of Pb(II). You might have commented that derivative polarograms would marginally improve the situation. In the next section (1.7) there is a hint of a solution to this problem.

****

SAQ 1.6c

A series of standard solutions of $Zn(NO_3)_2$ were prepared and their polarograms were obtained separately at 25 °C. A solution containing Zn(II) ions at a concentration of approximately 10^{-4} mol dm^{-3} was now investigated (the unknown). The results are given below. Determine the accurate concentration of Zn(II) ions in the latter solution.

$10^4 \times c(Zn^{2+})/0.50$ 2.00 3.00 5.00 unknown
mol dm^{-3}

$I_d/\mu A.$ 0.65 2.58 3.86 6.46   4.75

Suppose the unknown had been a sample derived from a sewage effluent which contains many species other than Zn(II). Do you consider the method that you have just used to be satisfactory?

Response

The answer is 3.7×10^{-4} mol dm^{-3}. You should have used the direct calibration (or working) curve method (Fig. 1.6j.)

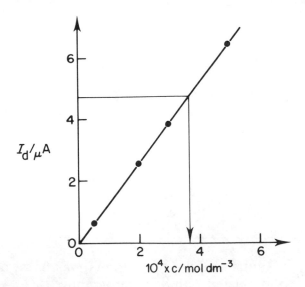

Fig 1.6j. *Application of a calibration curve*

The sewage sample would contain many metal cations and organic species. There would therefore be potentially many interferences, ie contributions other than Zn(II) ions to the Faradaic current. You could elect to clean up the sample using separation techniques and then use the direct method but it would be better to use the standard addition method.

****

SAQ 1.6d	Why is dc polarography said to be a non-destructive method of analysis? Name one other non-destructive and one destructive method of analysis.

Response

Each experiment typically removes $<0.5\%$ of the analyte from solution – hence virtually non-destructive. Each student will have a different background knowledge but examples of non-destructive techniques are: uv-visible spectrometry, infrared spectrometry; examples of destructive techniques are: atomic absorption spectroscopy, flame photometry.

****

SAQ 1.7a   Summarise the main limitations and problems associated with dc polarography.

Response

You should have included the following points.

(*i*)   The finite voltage window which at best is $+0.2$ to -3.0 V (SCE).

(*ii*) The detection limit is about 10^{-4} to 10^{-5} mol dm^{-3} which is >1 ppm.

(*iii*) Overlap of waves from species with similar values of $E_{\frac{1}{2}}$.

(*iv*) The existence of three specific effects which cause distortion of the wave or the emergence of additional waves. These are catalytic currents, adsorption currents and kinetic currents and their effects are due to changes occurring in the normal model of the transport of the analyte to the electrode.

****

| SAQ 2.1a | What is meant by the term activation energy? |

Response

The energy needed to get the reaction started. At the beginning of a reaction an initial reaction intermediate, the so called activated complex, must be formed. The formation of this excited state requires energy, the activation energy. The activated complex decomposes to the products. In polarography the activation step would involve addition of an electron, possibly across some energy barrier. If the activation energy is small compared to the energy difference between starting materials and products, a reversible electrolytic process will occur. If the activation energy is large compared to the energy difference between starting materials and products, an irreversible electrolytic process will occur.

****

| SAQ 2.2a | What is Faraday's Law? |

Response

In modern language this states that the quantity of electricity required to make one mole of an electrode product is a universal constant multiplied by the number of electrons required per molecule (or per ion), the constant is the Faraday (symbol F) with a value of 96485 coulombs or 96485 ampere seconds. It could be expressed:

$$\text{No. of moles formed} = nFIt$$

where I is the current in amperes, t the time in seconds, n the number of electrons consumed per molecule ($=$ the charge on a metal ion) and F the Faraday. It could be summed up by saying that one mole of electrons is equivalent to one Faraday or 96485 coulombs.

****

SAQ 2.2b

> What effect would migration current have on the overall limiting current, or wave height, for (*i*) electrostatic attraction and (*ii*) repulsion, respectively, of the analyte to the electrode?

Response

The electrostatic attraction will increase the mass transport of ions to the electrode and will thus add to the current. The electrostatic repulsion will slow down the mass transport of ions to the electrode and will thus lower the current.

****

SAQ 2.2c

> From experimental data how might one determine to which power of the time the current is proportional?

Response

The simplest method is to plot log(current) against log(time). The slope gives the power to which the time is raised in the relationship. This is most accurate near the end of the drop lifetime. The same method could be used for the relationship between the current and the mercury column height. It is a good idea to replot the current against the power of the time (or column height) found to check that the line passes through the origin (0,0).

****

SAQ 2.2d   What is the essential difference between a catalytically and a kinetically controlled wave?

Response

The key difference lies in the fact that the species of interest is regenerated and not consumed in the overall electrode process of the catalytic wave. In the kinetic wave the species of interest is consumed by the preceding chemical reaction. Hence catalytic waves are very large and conventional kinetic waves small. Strictly speaking a catalytic wave is a kinetic wave with regeneration of the species of interest. But usually the term kinetic is applied to processes without regeneration.

****

SAQ 2.2e   The height of a dc polarographic wave was measured at different heights of the mercury column for a freely dropping mercury electrode. The following wave heights were obtained, for these mercury column heights corrected for back pressure.

height of column/cm	30	40	50	60	70
wave height/arbitrary units		23.5	27.3	32.0	33.5	36.2

$\longrightarrow$

SAQ 2.2e
(cont.)

> A good proportional relationship was found for the wave height with concentration. The wave height was found to increase with temperature by about 2% K^{-1}. Which type of limiting current was involved?

Response

A plot of log(wave height) against log(Hg-height) should have given a slope of about 0.5. This is a characteristic of diffusion control. The other data also points to diffusion control. The wave probably will be suitable for analytical application.

****

SAQ 2.2f

> A substance of interest gave rise to a wave whose height was independent of the height of the mercury column. The wave height was proportional to the concentration and increased with temperature by about 15% K^{-1}. Which type of limiting current was involved?

Response

The effect of temperature indicates kinetic character. The other data is consistent with complete kinetic control by a preceding first order chemical reaction. A catalytic process could not be ruled out. (Continued electrolysis on a very small volume at the potential of the upper plateau, would show whether the analyte was consumed or regenerated). Either way the wave would generally not be suitable for analytical application.

****

SAQ 2.2g The dc polarographic wave height was found to be proportional to the concentration of the analyte. The wave height was further found to be proportional to the square root of the height of mercury column while the current during the drop lifetime was found to be proportional to $t^{0.23}$, where t is the time from the birth of the drop. The wave height increased by 12% K^{-1} with rising temperature. Which type of limiting current was involved?

Response

This doesn't exactly fit any of the diagnostic types described in this section does it? Real cases often cannot be classified into water-tight theoretical boxes. Apart from the effect of temperature this example would appear typical of diffusion-control. But the large increase of wave height with temperature could not be due to an increase in the diffusion rate. This alone indicates a kinetic character, although not of the simple type described. The high temperature coefficient would render the wave unsuitable for analytical application where high precision is required.

****

SAQ 2.2h

The height of a dc polarographic wave was measured at different heights of the mercury column for a freely dropping mercury electrode. The following wave heights were obtained, for these mercury column heights corrected for back pressure.

height of column/cm	30	40	50	60	70
wave height/arbitrary units		15.7	17.3	18.0	19.2	20.0

A reasonably proportional relationship was found for the wave height with concentration. The wave height was found to increase with temperature by about 7% K^{-1}. Which type of limiting current was involved?

Response

The log (wave-height) against log (Hg-height) plot has a slope about 0.3, that is half-way between the value for full kinetic control and full diffusion control. This most likely is a diffusion-controlled wave with a kinetic component. The temperature coefficient is not drastically high. An analytical application might be possible but with a distinctly reduced precision.

****

SAQ 2.2i | The main polarographic wave of interest was preceded by a small prewave. The height of this prewave was independent of the concentration of the analyte. The wave height of the prewave was found to be proportional to the square root of the height of mercury column while the current during the drop lifetime was found to be proportional to $t^{0.23}$, where t is the time from birth of the drop. The height of the prewave increased only to a small extent with rising temperature. What could be the origin of this prewave?

Response

A small wave of constant height at the base of the main wave sounds exactly like an adsorption prewave. But remove the main wave and the characteristics of this small wave fulfil the conditions for a diffusion-controlled wave for a substance at a fixed concentration. This small wave is probably due to an electroactive impurity in the supporting electrolyte. The running of a blank on the supporting electrolyte would show this, unless the impurity is rendered electroactive only by reaction with the analyte. This example shows that the mercury column height or time dependency of the wave height should always be investigated.

****

SAQ 2.2j | Which are the only types of current to decrease during the drop lifetime?

Response

Adsorption currents and capacitive currents. Both are dependent on the amount of freshly created surface on the growing drop. Towards the end of the drop lifetime the drop grows more slowly and there is less and less fresh surface for adsorption or requiring charge to reach the set voltage. Thus adsorption and capacitive currents fall with time. All the other currents are increased by having a larger electrode. A current which rises and falls during the drop lifetime indicates complex adsorption phenomenon such as inhibition effects.

****

SAQ 2.2k   Why is it best to avoid the region around pK values for polarographic determination of an analyte?

Response

Often the current falls near the pK value as the electroactive form protonates or deprotonates. This would, of course, cause a small loss in sensitivity. But far more important is the fact that the wave would probably develop a kinetic component and lose its reproducibility due to increased temperature sensitivity.

****

SAQ 2.3a   If the wave height is independent of pH, why should it be necessary to buffer samples and standards at the same pH?

Response

It is always a good principle to keep the standards as close as possible in character to the expected samples. Also in practice, when the wave height is found to be independent of pH, there is in fact usually a small drift in the height with pH. This would not be significant from a physical chemical understanding of the mechanism, but would be a significant source of error in precise analysis. The limiting current itself might be pH independent but its measurement as a wave height construction could be slightly pH dependent.

****

SAQ 2.3b	Would you expect the potential of the reduction or oxidation of the supporting electrolyte itself to shift with pH?

Response

Of course, typically the decomposition of the supporting electrolyte is reduction of the hydrogen ion or oxidation or reduction of water or of the major ions present such as sulphate etc. Decomposition of water or hydrogen ion obviously involves hydrogen ion, as does the decomposition of many common buffer components and so the potential of the supporting electrolyte is pH-dependent. The cathodic (reduction) potential range available in acid is much smaller than in most neutral solutions.

****

SAQ 2.4a	What effect would aggregation of analyte molecules be likely to have on the polarographic behaviour of that analyte?

Response

The aggregation or 'clumping together' of molecules will obviously alter the rate of diffusion through the solution. This will usually result in a lessening of the wave height. The aggregation could also alter the potential of the electrode process if energy were required to break up the aggregation before the electrolysis. This sometimes happens sometimes not. Aggregation will certainly disturb the calibration.

****

SAQ 2.4b	How might an organic solvent alter the adsorption behaviour of an analyte?

Response

Adsorption is an equilibrium between an affinity for the surface and an affinity for the solution. An organic solvent by altering the state of solvation can often alter this balance.

****

SAQ 2.5a	If the introduction of a hydroxy substituent causes only a very small shift of the half wave potential of a particular compound, does this mean that a method, using polarography, cannot be developed to selectively determine both the parent compound and its hydroxy derivative?

Response

The key phrase is a method using polarography. The hydroxy group would alter the acid-base chemistry of the compound. This factor could be used in a preceding separation step, such as solvent extraction, to separate the two compounds. Polarography could still be used for the quantitative detection or determination step, and thus offer selective determination of both forms independently.

****

SAQ 2.6a	If several different complexing agents are present in the solution, how will this affect the polarographic behaviour?

Response

With most metal ions the strongest available complex is always formed. Other weaker complexing agents will have no effect unless present in very large excess. Of course each metal ion could 'choose' a different complexing agent. The only complexing agents to interfere are those that are stronger than the one originally chosen.

****

SAQ 2.6b	If two metallic ions have similar half wave potentials as the free aquo-ion and both form complexes with a given complexing agent, would you expect the two ions to have similar half wave potentials in the presence of an equal excess of that complexing agent?

Response

Generally no. The two aquo-ions might have similar half wave potentials but their complexes, with the given ligand, need not have the same stability constants. If they do another complexing agent can usually be found to separate them. Obviously if the two metal ions are very similar in their overall chemistry then it may be very difficult to find a complexing agent to separate their half wave potentials.

****

SAQ 2.7a	What is the major effect of depletion of the analyte on the voltammetric signal?

Response

The current signal depends on the rate at which the analyte reaches the electrode surface. Depletion means that this must come from further and further out, thus reducing the rate at which it reaches the electrode thus reducing the current signal. Stirring the solution reduces the depth of this depletion or diffusion zone and allows more rapid transport to the electrode surface, thus restoring the current signal.

****

SAQ 2.7b	Why should a fast scan rate produce a large charging or capacitive current at a solid electrode?

Response

When the voltage increases by a certain amount, a fixed increase in the charge density on the electrode surface must occur. If the scan rate is low this amount of charge is added slowly, giving rise to a small current. If the scan rate is fast the same charge in coulombs must be added in a much shorter time and so a large charging current in coulombs/second (amperes) must flow. If the scan rate is in terms of several volts per second this charging current is considerable.

****

SAQ 2.7c
> How also might the reversibility of an electrode process be investigated?

Response

The variation of the half wave potential with pH also yields information about the reversibility of the electrode process (2.2). The slope of the plot of half wave potential against pH for a reversible process will be 0.059 p/n V/pH unit, (n is usually 1). That is for a reversible process the slope will be a simple ratio or multiple of 0.059 V. For an irreversible process the slope is given by 0.059 $p\alpha n$ V/pH unit, that is no longer a simple ratio or multiple of 0.059 V since the transfer coefficient α has a value significantly less than 1.

Note that this use of the half wave potential/pH relationship reveals the apparent reversibility of the potential-determining step only while cyclic voltammetry gives information on the reversibility of the overall electrolysis process. (For some compounds the initial potential-determining step can appear to be reversible while the overall process is irreversible).

****

SAQ 2.7d   Why must the electrode surface be regularly pol-
                      ished?

Response

Electrode products can often build up on the surface of a solid
or stationary electrode. Species from the solution may adsorb on
the surface. Some solid electrode materials may even develop oxide
layers on the surface. All of these produce changes in the nature
of the electrode and often completely distort the electrochemical
response. By regular polishing back to a clean metal or graphite
surface, some attempt can be made to achieve a reproducible surface. Better results are usually obtained from a smooth rather than
a rough surface.

****

SAQ 3.2a   Sum up, in two key words, the factors limiting
                      the sensitivity of dc polarography.

Response

Depletion, Capacitive Current.

If you did not choose these two keywords you have not fully under-
stood this section so read it carefully again with these two words in
mind.

****

SAQ 3.3a	For a drop lifetime of 2 seconds, what fraction of the drop lifetime is taken by the pulse?

Response

2.8%, hence the great reduction in the depletion of the electroactive species in normal pulse polarography.

****

SAQ 3.3b	Why does the normal pulse polarographic wave in Fig. 3.3b not display the rapid current fluctations found with the classical dc wave?

Response

The rapid fluctuations in current with classical dc polarography occur because the current is measured throughout the drop lifetime and increases rapidly as the drop grows but falling to zero as the drops falls. The fluctuations require to be reduced towards the mean by means of an electronic damping circuit. In normal pulse polarography the current is measured in so short a pulse that the electrode is almost of constant size, during the pulse, and so fluctuations due to growth are almost eliminated.

****

SAQ 3.3c   If the potential pulse is to occur at the exact same point in the drop lifetime, how might the pulse be synchronised to the dropping of the mercury electrode?

Response

The drop lifetime must be kept to a chosen value. This is most easily done by knocking the drop off, with an electronically controlled hammer, after a fixed chosen time interval. The potential pulse signals can then be synchronised to the hammer. The natural drop lifetime must naturally be longer than that enforced by the hammer.

****

SAQ 3.4a   Summarise in one sentence each, for both methods, the potential signal imposed and the current signal measured for normal and differential pulse polarography.

Response

Normal pulse polarography – a series of very short potential pulses, one towards the end of each drop lifetime, rising in a linear ramp is imposed with the potential returning to a base value in between the pulses; the current is measured in an even shorter pulse towards the end of each of the potential pulses.

Differential pulse polarography -- on to a conventional rising linear voltage ramp is imposed a very short potential pulse of small fixed amplitude, one pulse towards the end of each drop lifetime; the current is measured in two even shorter pulses just before and just at the end of this potential pulse.

O.K. you didn't make it in one sentence each. The idea of the exercise is to make you bring it all together and by summarising fix the overall shape in your mind. If you left out anything mentioned above reread the appropriate section carefully.

∗∗∗∗∗∗∗∗∗∗∗∗∗∗∗∗∗∗∗∗∗∗∗∗∗∗∗∗∗∗∗∗∗∗∗∗

SAQ 3.4b
Is the sensitivity of the recording apparatus likely to be of importance to the overall sensitivity of the polarographic method?

Response

With todays instrumentation the only major source of noise comes from the electrolysis cell itself. The higher overall currents with normal pulse polarography in themselves offer no advantage. This is because the sensitivity of the electronic components is many orders of magnitude better, and the noise many orders lower, than that of the electrolysis cell. Thus although differential pulse polarography produces a smaller signal, its better resolution of Faradaic and capacitive current gives it the higher sensitivity.

∗∗∗∗∗∗∗∗∗∗∗∗∗∗∗∗∗∗∗∗∗∗∗∗∗∗∗∗∗∗∗∗∗∗∗∗

SAQ 3.5a
What is the equivalent equation for classical dc polarography?

Response

The Ilkovic equation: $I_d = 708 \, nD^{\frac{1}{2}} \, m^{\frac{2}{3}} \, t^{\frac{1}{6}} c$ (1.5d)

Note that the Ilkovic equation contains a term m, the mercury flowrate, which reflects the growth of the drop during the measurement process. In contrast the Parry-Osteryoung equation contains a term A for the surface area of the electrode as if this were a constant. A is for practical purposes a constant since the growth of the drop is negligible during the very short pulse.

****

SAQ 3.5b	What is meant by thermodynamically reversible or irreversible?

Response

Thermodynamical reversibility in the electrochemical context is concerned with the kinetics or rates of the electrode reactions and whether the electrochemical reactions occur in equilibrium. Reversible reactions involve a process in which forward and back (reverse) reactions are both rapid and remain in equilibrium. The overall reaction is a balance of forward and back. In an irreversible reaction the back reaction is far too slow to be significant and only the forward reaction is considered. The process occurs out of equilibrium.

If this is too difficult to understand, do not worry about it. The important facts to remember are: (i) if a electrode reaction is reported to be irreversible the polarographic wave or peak will often be less well formed than for a reversible process; (ii) irreversibility causes classical dc and normal pulse polarographic waves to be less steep but has little effect on their height; (iii) on the otherhand irreversibility broadens and lowers the height of the differential pulse polarographic peak but the area under the peak is little affected.

****

| SAQ 3.5c | What methods could be used to actually measure the area under the differential pulse peak on the chart recorder output? |

Response

Several methods are possible. Counting of squares under the curve on the chart paper is one method. The accuracy would depend on the size of the squares. Another method would be to cut out the paper under the peak and weigh it. The weight of a known square of the same paper would give a calibration. A high qual ity paper of proven uniform density is required. Various mechanical and electronic devices exist to measure areas under curves. The more expensive polarographic instruments are fitted with integrating circuits, particularly those with built-in microprocessor interfacing. This does not apply to the cheaper instruments favoured in many laboratories, because of their cost. It is best to try several of any methods available and use that which gives the most reproducible calibration with standards.

All the methods share a common problem – the need to identify the base line below the curve. One possibility is to 'run' a curve on a 'blank' solution containing none of the analyte, and use this as a baseline. However a real solution containing no analyte may not yield the same result as the theoretical result obtained by extrapolation of several curves to a theoretical zero analyte concentration. It is possible to extrapolate the flat plateau on either side of the peak to form a baseline. But, the true baseline might not be so flat! Other methods contain assumptions. It is best to try several methods and use that which gives the most reproducible calibration with standards. This will all be dealt with later.

****

SAQ 4.2a	Why would the HMDE and the DME be unsuitable for use in such flow systems?

Response

They are much too mechanically unstable. The drop can be knocked off far too easily. They have in fact been used in flow systems but great care is needed and much patience. Further it is difficult to construct very small cells to hold a mercury drop electrode – a major disadvantage.

****

SAQ 4.2b	A 20.0 cm^3 sample of effluent contains approximately 10^{-8} mol dm^{-3} of copper ions. At the chosen electrode this generates about 1 nA of current during the deposition step. What length of deposition time would be required to deposit about 2% of the copper?

Response

The number of moles of Cu^{2+} in the sample

$$= 20.0 \times 10^{-3} \times 10^{-8}$$
$$= 2 \times 10^{-10}$$

The number of moles of Cu^{2+} deposited $= 0.02 \times 2 \times 10^{-10}$
$$= 4 \times 10^{-12}$$

The number of moles of electrons required $= 8 \times 10^{-12}$

The number of coulombs required $= 96485 \times 8 \times 10^{-12}$
$$= 7.72 \times 10^{-7}$$

Thus time required $= 7.72 \times 10^{-7}/1 \times 10^{-9}$ s
$$= 772 \text{ s}$$
$$= 12.9 \text{ min}$$

Thus about 15 minutes deposition time would be suitable. A current of 1 nA flowing in the deposition would be extremely difficult to measure. It would in fact be overwhelmed by the background current. However during the stripping step the same number of coulombs 7.72×10^{-7} required to deposit the copper must pass in the other direction. The stripping peak is about 200 mV wide and at a scan rate of 200 mV s^{-1} this means that the same number of coulombs are now passed in about one second. The average current in the stripping step becomes 7.72×10^{-7} A with a peak height somewhere about 1×10^{-6} A. This is very much easier to measure. You can see in fact that the increase in sensitivity is roughly the ratio of the long deposition time to the short stripping time.

****

SAQ 4.3a
What is usually meant by the terms anodic and cathodic and when is an electrode an anode or a cathode?

Response

An anodic reaction is an oxidation. A cathodic reaction is a reduction. An anode is an electrode at which an oxidation reaction is taking place. Its potential need not have a positive value but oxidation occurs more easily the more positive the potential. Thus the anodic direction is to more positive potentials. Similarly a cathodic or reduction process could occur at a positive potential but as the electrode becomes more negative it becomes more cathodic.

****

SAQ 4.3b   What is meant by the terms Faradaic current and capacitive current?

Response

A Faradaic current is one caused by some kind of electrochemical process and is related to species in the solution. The capacitive current is the current required to charge the electrode up to the chosen potential.

****

SAQ 4.3c   Why would an unstirred solution offer greater reproducibilty?

Response

Simple undisturbed diffusion is a highly reproducible phenomenon and highly suitable for an analytical application. In a stirred solution the movement of the solution can be highly complex and not always predictable. If eddies form in the stirred solution they are very unreproducible and must be avoided. Even in a very evenly stirred solution a slight displacement of the electrode can change the flow pattern. Reproducible stirring requires careful cell design and experimentation. The best stirring is with a rotating electrode in a reproducible position in the cell.

****

| SAQ 4.3d | Convert this range of typical detection limits 10^{-10} to 10^{-9} mol dm^{-3} to ppb for a first period transition metal ($A_r = 50$). |

Response

10^{-10} mol dm^{-3} is equivalent to 50×10^{-10} g dm^{-3}

$$= 5 \times 10^{-9} \text{ g dm}^{-3}$$

$$= 0.005 \ \mu\text{g dm}^{-3}$$

$$= 0.005 \text{ ppb}$$

Thus the limit of detection would be 0.005–0.050 ppb. This is in fact an almost meaningless figure since it is usually impossible to obtain reagents of this level of purity. The stripping voltammetric step often has an intrinsic sensitivity greater than that of the complete method of which it will be part. Of course the limit of detection in an individual case can vary considerably.

****

| SAQ 4.4a | How would you find out if a second metallic ion in the sample would cause interference by forming an intermetallic compound with the analyte? |

Response

The literature might give a warning but ultimately this can only be checked by experimental trial. A number of different standard concentrations of the analyte would be chosen. The potential interfering metal ion would be increased stepwise in these, to see if this altered the proportionality of the calibration for the analyte or shifted its peak. The method would only be used where no interference was detected. When a method from the literature is being used, interferences and levels of interference from the literature should not be taken on trust but tested if possible. They are not always reproducible in the new environment.

****

SAQ 4.4b   In one sentence state why stripping voltammetry is the most sensitive of all voltammetric methods?

Response

It is the most sensitive because it includes a preconcentration step in which the analyte is collected over a long period from a large volume and reoxidised (or re-reduced) in a short 'burst' creating an enhanced current signal.

****

SAQ 4.4c   In a given electrolyte Fe(III) gives rise to two dc polarographic cathodic waves. On the upper plateau of which wave would the deposition potential be chosen for an anodic striping voltammetric determination of Fe(III)?

Response

Iron(III) is reduced in two stages, first to iron(II) and then on to the metallic form. Depending on conditions this could occur as one single wave or as two. If two waves are formed the more positive would be reduction to iron(II) while the more negative would be the reduction of this iron(II) to the metal. The product of the first, more positive wave, iron(II) is soluble. Thus this wave could not be used for stripping voltammetry. The second, more negative, wave has a metallic product, therefore stripping voltammetry would employ a deposition potential on the upper plateau of this second wave.

****

SAQ 4.4d
Could Fe(II) and Fe(III) be differentiated by stripping voltammetry in the electrolyte mentioned in SAQ 4.4c?

Response

No. As iron(II) is the product of the first wave, this first wave would not be present if only iron(II) were present. In a mixture of iron(II) and iron(III) the height of the first dc polarographic wave would give the concentration of iron(III). The second wave height would depend on the original iron(II) present plus the iron(II) formed by reduction of iron(III) in the first wave. That is the height of the second wave will give the total iron in the solution. Dc polarography can thus selectively determine the two species. However reduction at the potential of the upper plateau of the second wave will deposit both iron(III) and iron(II) to the metallic form and thus stripping voltammetry will not differentiate the two forms. The first wave does not yield a suitable insoluble product and cannot be used. Since dc polarography or differential pulse polarography could differentiate the two forms quantitatively, stripping voltammetry would only be of interest if the sensitivity of the direct polarographic methods was insufficient.

****

SAQ 4.4e   A sample of saline effluent was divided to give
two 20.0 cm^3 aliquots. One of these was subjected to anodic stripping voltammetry. A deposition of 10 minutes at -0.8 V was chosen. A
subsequent anodic stripping peak, of 24.6 units
in height on the chart paper, was found to correspond to cadmium. To the second 20.0 cm^3
aliquot was added 0.1 cm^3 of a standard solution of 5×10^{-6} mol dm^{-3} of cadmium ions.
This gave a new anodic stripping peak for cadmium of 39.8 units under identical conditions.
Calculate the concentration of cadmium ions in
the effluent in ppb. $[A_r(Cd) = 112.4]$

Response

The number of moles of standard cadmium added

$$= 0.1 \times 10^{-3} \times 5 \times 10^{-6}$$
$$= 5 \times 10^{-10} \text{ mol}$$

This is responsible for the increase in peak height of $39.8 - 24.6$
units

$$= 15.2 \text{ units}$$

Therefore one unit is equivalent to $5 \times 10^{-10}/15.2$ mol

$$= 3.29 \times 10^{-11} \text{ mol}$$

The original aliquot produced a peak height of 24.6 units

The original aliquot therefore contained $24.6 \times 3.29 \times 10^{-11}$ mol

$$= 8.09 \times 10^{-10} \text{ mol}$$

This was in a sample volume of 20 cm^3 or 0.020 dm^{-3}

The concentration of the cadmium in the sample

$$= 8.09 \times 10^{-10}/0.020 \text{ mol dm}^{-3}$$
$$= 4.05 \times 10^{-8} \text{ mol dm}^{-3}$$

The relative atomic mass of cadmium is 112.4

The concentration of cadmium in the effluent is therefore

$$= 112.4 \times 4.05 \times 10^{-8} \text{ g dm}^{-3}$$
$$= 4.55 \times 10^{-6} \text{ g dm}^{-3}$$
$$= 4.55 \text{ } \mu\text{g dm}^{-3}$$
$$= 4.55 \text{ ppb}$$

The above calculation contains two dangerous assumptions. It assumes a linear calibration of peak height with concentration. This could be checked by making several such standard additions. It also assumes that the reagent contain no cadmium traces. A blank would have to be run on any reagents used, to check and allow for this.

****

SAQ 5.2a

> The EEC allowed limit for cadmium ion in drinking water is 0.050 ppb. Briefly consider the general issues on which you might base a decision whether to purchase atomic spectroscopy equipment or polarographic equipment for the determination of cadmium at this level.

Response

It is of course impossible for you to give a complete answer to this question. For a start you don't have any exact information about the types of drinking water involved. However some points could be considered. Cadmium has only one ionic oxidation state. The selectivity of polarographic methods to different oxidation states is not relevant here. In this respect the two methods are equal. A level of 0.050 ppb equivalent to 4×10^{-10} mol dm^{-3} is an extremely low level. It is too low for a direct measurement by either differential pulse polarography or flameless atomic absorption spectroscopy. With these a preconcentration step would be necessary. Anodic stripping voltammetry does have a sufficiently low limit of detection although this would be at the limit of its range. An anodic stripping method would certainly be worth exploring. The rate of required sample throughput might well be the important factor. If this were high the much more rapid atomic spectroscopic method might be preferred. (The time per sample would of course have to include the time for the preconcentration step). On the other hand if the laboratory were short of funds the cost of the much cheaper polarographic equipment might be the critical factor. Of course in practice other measurements would have to be made by the laboratory and these would have to be taken into consideration as well, in the purchase of equipment. The experience of laboratory staff would also be a factor. Finally the exact nature of the samples would have to be considered and a method carefully developed for the individual circumstances of the laboratory. The literature will contain many methods for this problem involving both polarography and atomic spectroscopy. These would have to be carefully read.

You probably won't have thought of all of these points but you should have thought of some of them. The question was not designed to test but to provoke some thought.

****

SAQ 5.2b   Why can stripping voltammetry be applied to only a very small number of organic compounds?

Response

Most organic compounds do not form insoluble electrode products. Most organic compounds are soluble at these high dilution trace levels. Many of those organic electrode products that are insoluble cannot be reoxidised or reduced to form a stripping peak, or the insoluble layer is insufficiently coherent. The most common type of organic compound to which stripping analysis can be applied are those containing a thiol group.

****

SAQ 5.3a	What are the three most important roles played by the separation step?

Response

(*i*) To separate the analyte from a medium in which it cannot be determined.

(*ii*) To remove potential interfering species.

(*iii*) To separate and allow individual determination of electro-chemically similar analytes.

****

SAQ 5.3b	In the above example the selectivity between the species of interest is obtained in the solvent extraction steps. What might be the advantage in using polarography for the final determination step?

Response

The solvent extraction system has been designed primarily to differentiate the different phenothiazine derivatives and to remove material that could interfere with the polarographic step. The final solutions subjected to polarography are still likely to be highly complex. The advantage of the polarography is that in these solutions the drug metabolites are the only electroactive compounds at these potentials. Another technique of similar sensitivity to differential pulse polarography might well find the extracts contain too many interfering compounds. In reality the solvent extraction and polarographic steps are designed as an overall process with the steps in harmony. The polarographic step offers its own selectivity, perhaps not between the metabolites of interest, but between them as a class and other extracted substances. The overall selectivity of the method comes from both the solvent extraction and polarographic steps.

****

SAQ 5.6a	What is the difference between accuracy and precision?

Response

Precision is a measure of the random scatter of the analytical results about a mean value. It is a measure of the reproducibility of the results. It gives no information as to how far from the 'true' value the mean analytical value is. Accuracy is the measure of how far from the 'true' value the mean analytical value is. A constant difference between mean analytical and 'true' values is known as a systematic error.

In many trace analytical problems the 'true' value may never be known. Ultimately in analytical chemistry something must be chosen as a standard and assumed to have a 'true' value. In trace analysis the search for this standard becomes more difficult as the levels concerned become smaller. Systematic error problems also become more difficult at lower concentration levels. This is a problem you will meet professionally. It is a matter of experience which you will gain in practice.

****

SAQ 5.6b	The preparation of a calibration plot and of a sample measurement involves the running of a 'blank'. What is a 'blank'?

Response

A blank is a measurement taken in the absence of the analyte. It will be taken, for example, to determine traces of the analyte (or material which could be mistaken for it) in the reagents used, such as the supporting electrolyte. This is particularly important in trace metal analysis. Stripping voltammetry can for example easily detect the metallic ion impurities present in AnalaR reagents, or even in distilled water. Another example is in the polarographic analysis of a drug, or its metabolites, in blood plasma, a suitable blank would be a blood plasma sample from the patient before the drug is administered. This would indicate if any interfering electroactive materials were carried across in the extraction process. A blank is the zero position in a calibration plot and must be determined in order to remove systematic errors and permit a proportional relationship with concentration to be established. A problem in trace analysis is to produce a sample known to contain zero analyte.

****

SAQ 5.6c   How in fact would you determine which of the two calibration methods to use for a particular analytical problem?

Response

A general consideration of the variability of the sample matrix would be a guide. There is however no substitute for trying it out experimentally. The two calibration methods should in fact be used on a representative set of standard samples and both their precision and accuracy noted. The best technique should then be chosen. If there is only a small number of samples, the standard addition technique would often be more rapid.

****

Units of Measurement

For historic reasons a number of different units of measurement have evolved to express quantity of the same thing. In the 1960s, many international scientific bodies recommended the standardisation of names and symbols and the adoption universally of a coherent set of units—the SI units (Système Internationale d'Unités)—based on the definition of five basic units: metre (m); kilogram (kg); second (s); ampere (A); mole (mol); and candela (cd).

The earlier literature references and some of the older text books, naturally use the older units. Even now many practicing scientists have not adopted the SI unit as their working unit. It is therefore necessary to know of the older units and be able to interconvert with SI units.

In this series of texts SI units are used as standard practice. However in areas of activity where their use has not become general practice, eg biologically based laboratories, the earlier defined units are used. This is explained in the study guide to each unit.

Table 1 shows some symbols and abbreviations commonly used in analytical chemistry; Table 5 is a glossary of abbreviations used in this particular text. Table 2 shows some of the alternative methods for expressing the values of physical quantities and the relationship to the value in SI units.

More details and definition of other units may be found in the *Manual of Symbols and Terminology for Physicochemical Quantities and Units*, Whiffen, 1979, Pergamon Press.

Table 1   *Symbols and Abbreviations Commonly used in Analytical Chemistry*

Å	Angstrom
$A_r(X)$	relative atomic mass of X
A	ampere
E or U	energy
G	Gibbs free energy (function)
H	enthalpy
J	joule
K	kelvin $(273.15 + t\,°C)$
K	equilibrium constant (with subscripts p, c, therm etc.)
K_a, K_b	acid and base ionisation constants
$M_r(X)$	relative molecular mass of X
N	newton (SI unit of force)
P	total pressure
s	standard deviation
T	temperature/K
V	volume
V	volt $(J\ A^{-1}\ s^{-1})$
$a, a(A)$	activity, activity of A
c	concentration/ mol dm^{-3}
e	electron
g	gramme
i	current
s	second
t	temperature / °C
bp	boiling point
fp	freezing point
mp	melting point
$\approx$	approximately equal to
$<$	less than
$>$	greater than
e, $\exp(x)$	exponential of x
$\ln x$	natural logarithm of x; $\ln x = 2.303 \log x$
$\log x$	common logarithm of x to base 10

Table 2 *Alternative Methods of Expressing Various Physical Quantities*

1. **Mass (SI unit : kg)**

$$g = 10^{-3} \text{ kg}$$
$$mg = 10^{-3} \text{ g} = 10^{-6} \text{ kg}$$
$$\mu g = 10^{-6} \text{ g} = 10^{-9} \text{ kg}$$

2. **Length (SI unit : m)**

$$cm = 10^{-2} \text{ m}$$
$$\text{Å} = 10^{-10} \text{ m}$$
$$nm = 10^{-9} \text{ m} = 10\text{Å}$$
$$pm = 10^{-12} \text{ m} = 10^{-2} \text{ Å}$$

3. **Volume (SI unit : m³)**

$$l = dm^3 = 10^{-3} \text{ m}^3$$
$$ml = cm^3 = 10^{-6} \text{ m}^3$$
$$\mu l = 10^{-3} \text{ cm}^3$$

4. **Concentration (SI units : mol m^{-3})**

$$M = \text{mol } l^{-1} = \text{mol dm}^{-3} = 10^3 \text{ mol m}^{-3}$$
$$\text{mg } l^{-1} = \mu g \text{ cm}^{-3} = ppm = 10^{-3} \text{ g dm}^{-3}$$
$$\mu g \text{ g}^{-1} = ppm = 10^{-6} \text{ g g}^{-1}$$
$$\text{ng cm}^{-3} = 10^{-6} \text{ g dm}^{-3}$$
$$\text{ng dm}^{-3} = \text{pg cm}^{-3}$$
$$\text{pg g}^{-1} = ppb = 10^{-12} \text{ g g}^{-1}$$
$$\text{mg\%} = 10^{-2} \text{ g dm}^{-3}$$
$$\mu g\% = 10^{-5} \text{ g dm}^{-3}$$

5. **Pressure (SI unit : N m^{-2} = kg m^{-1} s^{-2})**

$$Pa = Nm^{-2}$$
$$\text{atmos} = 101\ 325 \text{ N m}^{-2}$$
$$\text{bar} = 10^5 \text{ N m}^{-2}$$
$$\text{torr} = mmHg = 133.322 \text{ N m}^{-2}$$

6. **Energy (SI unit : J = kg m² s^{-2})**

$$\text{cal} = 4.184 \text{ J}$$
$$\text{erg} = 10^{-7} \text{ J}$$
$$\text{eV} = 1.602 \times 10^{-19} \text{ J}$$

Table 3   *Prefixes for SI Units*

Fraction	Prefix	Symbol
10^{-1}	deci	d
10^{-2}	centi	c
10^{-3}	milli	m
10^{-6}	micro	μ
10^{-9}	nano	n
10^{-12}	pico	p
10^{-15}	femto	f
10^{-18}	atto	a

Multiple	Prefix	Symbol
10	deka	da
10^2	hecto	h
10^3	kilo	k
10^6	mega	M
10^9	giga	G
10^{12}	tera	T
10^{15}	peta	P
10^{18}	exa	E

Table 4 *Recommended Values of Physical Constants*

Physical constant	Symbol	Value
acceleration due to gravity	g	9.81 m s^{-2}
Avogadro constant	N_A	$6.022\ 05 \times 10^{23}$ mol^{-1}
Boltzmann constant	k	$1.380\ 66 \times 10^{-23}$ J K^{-1}
charge to mass ratio	e/m	$1.758\ 796 \times 10^{11}$ C kg^{-1}
electronic charge	e	$1.602\ 19 \times 10^{-19}$ C
Faraday constant	F	$9.648\ 46 \times 10^4$ C mol^{-1}
gas constant	R	8.314 J K^{-1} mol^{-1}
'ice-point' temperature	T_{ice}	273.150 K exactly
molar volume of ideal gas (stp)	V_m	$2.241\ 38 \times 10^{-2}$ m^3 mol^{-1}
permittivity of a vacuum	ϵ_0	$8.854\ 188 \times 10^{-12}$ kg^{-1} m^{-3} s^4 A^2 (F m^{-1})
Planck constant	h	$6.626\ 2 \times 10^{-34}$ J s
standard atmosphere pressure	p	$101\ 325$ N m^{-2} exactly
atomic mass unit	m_u	$1.660\ 566 \times 10^{-27}$ kg
speed of light in a vacuum	c	$2.997\ 925 \times 10^8$ m s^{-1}

Table 5 *Glossary and Abbreviations used in Electrochemistry*

A	area of electrode
C	coulomb
C_{DL}	double layer capacitance
D	diffusion coefficient
E	emf
$E(X^+,X)$	electrode potential of X^+,X
$E_{(PZC)}$	potential of zero charge
$E_{\frac{1}{2}}$	half-wave potential
F	Faraday constant
G	Gibbs (free) energy function
I	current
$I_a I_c$	anodic, cathodic current
I_{cp}	Capacitive current
I_d	diffusion current
I_{lim}	limiting current
I_m	migration current
J	cell constant
L	Length
Q	total charge
R	resistance
S	siemens
m	flow rate of mercury (mass/time)
n	number of electrons transferred
t	time

$t(X)$	transport number of X
$u(X)$	ionic mobility of X
z	charge number of ion
δ	thickness of electrical double layer
$\gamma_\pm$	mean ionic activity coefficient
ϵ_r	relative permittivity
η	overpotential, coefficient of viscosity
ρ	density
ω	ohm

Other abbreviations

AAS	Atomic Absorbtion Spectroscopy
DME	Dropping Mercury Electrode
DMF	dimethylformamide
DMSO	dimethylsulphoxide